NICHE AND ORGANIC CHICKEN PRODUCTS
Their technology and scientific principles

Cover design

Photographs by permission ~~of Poultry World~~

Niche and Organic Chicken Products

Their technology and scientific principles

SH Gordon and DR Charles

NOTTINGHAM
University Press

Nottingham University Press
Manor Farm, Main Street, Thrumpton
Nottingham, NG11 0AX, United Kingdom

NOTTINGHAM

First published 2002
© ADAS Consulting Ltd

British Library Cataloguing in Publication Data
Niche and Organic Chicken and Egg Products:
Their technology and scientific principles
I. SH Gordon. II. DR Charles

ISBN   1-897676-96-4

Disclaimer

Typeset by Nottingham University Press, Nottingham
Printed and bound by Hobbs the Printers, Totton, Hampshire, England

CONTENTS

INTRODUCTION

In developed countries modern consumers spend an ever-diminishing proportion of their income on food. Therefore, while there is of course still a substantial market for price driven commodity foods, there is also a market led by those who are looking for something more.

Nowadays food provides much more than merely nourishment and the elimination of hunger. It provides interest and enjoyment, and it has social and cultural associations. Furthermore consumers are increasingly interested in issues such as animal welfare, environment, provenance, special flavours, healthy eating and conforming with the standards set by special interest groups such as the organic certification bodies. Therefore niche poultry products are becoming increasingly important.

However, since producing them sometimes involves the application of techniques which have not hitherto been used in the production of commodity products, practical difficulties are emerging from which technical and scientific questions arise. This book addresses some of these difficulties and questions, and attempts to do so in an objective way leading to practical suggestions.

Most of the chapters are transcripts of reviews of the scientific literature commissioned by the Ministry of Agriculture, Fisheries and Food (MAFF), now the Department for Environment, Food and Rural Affairs (DEFRA), and adapted with permission. We gratefully acknowledge their financial support for these reviews. We particularly wish to mention the initiative and support of Dr. David Garwes, Head of Livestock Sciences. Since the chapters were originally separate pieces of work, some readers may perceive an element of discontinuity and some repititon, but we hope that the collection offers useful material on a range of topics relevant to niche production. There may be some virtue, for look-up purposes, in self-contained chapters.

In producing and editing these reviews we have been impressed with, though not particularly surprised by, the relevance to new problems of time honoured fundamental biological principles. This has led us to a general conclusion about the application of research to some of the modern niche issues.

Fundamental work, not originally aimed at niche production, addresses hypotheses on the ways in which farm animals and crop plants function, so that an understanding can be gained about how their needs may be met and how responses may be quantified. Response quantification can often permit economic appraisal. This means that reviewing literature, sometimes including quite old literature, can often offer useful guidance on new practical problems. However, by contrast, when studies of whole systems of production are required, then new research is often needed.

Before launching into the reviews of the scientific material we felt it appropriate to deal briefly with the historical background to the evolution of poultry production systems in the UK, and also to provide a brief analysis of the market background, with special reference to the UK market.

Some of the technical and marketing terms currently circulating are unclear. Thus there has sometimes been scope for ambiguity and lack of precision. In scientific work this is unacceptable. Therefore we have explicitly defined the terms used in this book.

The markets, the technology and the legislation affecting niche poultry products are all evolving fast. The information which we have reported in this book relates to the time of writing.

Finally we should like to put on record or thanks to the members of the research project workshops and steering groups. Their experience, wisdom and guidance has made progress possible. We are also grateful for the contributions on lighting by Peter Lewis, and on whole farm systems by Mark Shepherd and Anne Bhogul.

Sue Gordon
David Charles
Editors
June 2002

DEFINITIONS

Many of the niche poultry products are relatively new, and some of the necessary terminology has arisen rather quickly by the standards of the normal evolution of language. As a result some terms are in danger of becoming confusing. However clear definitions are needed before detailed chapters are attempted. Several of the definitions offered below were agreed by a study group on the production of organic pullets (see Chapter 9). They are marked * to distinguish them from definitions which are our own.

1. Production systems and product categories

DEEP LITTER

Originally, in the *traditional* period (see definition below), this term meant built up fermenting litter, often not renewed between batches of birds. A more appropriate term to describe the shallower layers of regularly renewed material, as generally used during the last 40 years, might simply be *litter*.

FREE RANGE

This term is used to simply mean poultry production systems in which the birds have continuous daytime access to pasture, though housing is invariably provided.

In the case of free range egg production, which is now such a large part of the British egg market that many consider it to be no longer a niche, minimum criteria came into force in 1985 under EC Regulation EEC 1943/85. These criteria specified stocking densities and building allowances among other things. Details of EC Regulations are not quoted here since they change frequently.

However there are now more specific criteria set by several of the retailers, so that the precise requirements and definitions depend upon the outlets through which the eggs are sold. In addition the Royal Society for the Protection of Animals (RSPCA) *Freedom Food* brand has its own standards to which members must conform (Freedom Food Ltd., Manor House, Causeway, Horsham, West Sussex RH12 1HG; 01403 264181).

FREE RANGE GRAZING AND MANAGEMENT SYSTEMS

1.   It is proposed that free range systems where the grazing moves around the whole farm, as a grass course in the rotation, might be termed *fully rotational free range, (FRFR)*.

2.   There are systems where the free range poultry are situated on the farm using the manure, or on a closely neighbouring farm, and where all the manure is utilised on the holding as a deliberate and planned component of the manurial policy. But the grass paddocks rotate around a fixed house in one location rather than moving round the whole farm. These might be called *rotationally integrated free range* systems, *(RIFR),* or simply *integrated free range (IFR)* systems for the sake of brevity.

*ORGANIC

This term is taken to mean conforming to the standards agreed by EC Directives, the UK Register of Organic Food Standards (UKROFS) and the certification bodies.

Note that this definition of the word differs from its original biochemical sense, meaning the chemistry of carbon compounds. Also it is of interest that Lampkin (1997) noted that the term *ecological products* might be more precise, but *organic* is now part of the general parlance of the food trade.

The specifications which determine that a product may be called *organic* originate from EC regulations. In UK these are overseen by UKROFS, which is the over arching certification authority under whose auspices several certification bodies operate. Examples of certification bodies include the Soil Association and Organic Farmers and Growers Ltd. There are others for Scotland and Ireland. Both UKROFS and the individual certification bodies have comprehensive documented specifications and rules for production to which producers must conform. However the spirit of all these rules is an aim to provide products which are land based and sustainable, with minimal reliance on inputs perceived as artificial, and with high standards of animal welfare and respect for the environment. For further details see the certification bodies and UKROFS, DEFRA, Nobel House, 17, Smith Square, London SW1P 3JR (020 7238 5605).

NICHE PRODUCTS

A *niche* was, in historical ecclesiastical architecture, a recess in a wall housing a special object, and in marketing the word implies speciality. Niches occupy small proportions of the market, unlike commodities. There is no consensus about how small the proportion is, but it is probably less than 5%. Niches are not fixed and they are not necessarily permanent. In expanding markets niche products often gradually become commodities.

2. Historical periods

Several definitions are needed to describe the phases of the history of poultry production in the UK.

*CONVENTIONAL

This term is appropriate to describe the poultry production technology used in the main stream UK commercial poultry industry since 1953/54 (see Chapter 1).

Note that the term *conventional* is often used of products to distinguish them from organic. In this book the term is reserved for the historical period: *non-organic* is used as the opposite of organic.

*TRADITIONAL

This term has often been used as a synonym for *conventional,* as defined above, but that is confusing. One of the study groups therefore agreed that it should be used to describe the technology of the period before the advent of the *conventional* period, *i.e.* before 1953/54.

*HISTORIC

For completeness the additional term *historic* is proposed to describe the poultry technology of the period before World War I.

3. The comparison of breeds

DEGREE OF MATURITY

This is a term due to Taylor (1980), and defined as the current liveweight of an animal divided by its mature weight.   He found that the growth of several species could be shown to fit on one curve if the *degree of maturity* was plotted against *metabolic age* (see definition below) rather than against chronological age.

GOMPERTZ CURVE

This is the name given to one of the most frequently used mathematical equations describing the sigmoid (S shaped) growth of animals, where weight is plotted against age or time. A review of the various equations used to describe the growth of chickens was given by Wilson (1977). Gompertz was an actuary who developed the equation in 1825. More detail of the biology and mathematics is given by Lawrence and Fowler (1997). Suffice it to say for the purpose of this definition that the equation has been found to describe the growth of animals so well that it has become conventional amongst those needing mathematical descriptions of biology. Growth rate accelerates at first, then slows as animals approach maturity.

METABOLIC AGE

Some niche markets require birds grown to older ages than has been usual in the *conventional* period. Breeds differ in their potential growth rates, so that comparisons of factors such as carcass quality at given ages, or at given liveweights, may be misleading. Therefore for use in applications such as meat quality we have adapted for poultry the application of the term *metabolic age,* θ, (Gordon *et al.*, 2001), as originally defined by Taylor (1965 and 1968).

$$\theta = t/A^{0.24}$$

where t = days from the start of incubation and A is mature weight, kg.
0.24 is a specific power for chickens published by Laird (1966)

4. Classes and types of chickens

BROILER

A *broiler* is a male or female hybrid chicken of a modern fast growing strain, bred by a specialist breeding company.

The word is probably derived from the verb *to broil*, which is often considered to be an American term for *to grill*. Yet the verb occasionally appears in English 19th century literature as a synonym for *to grill*. Broiler chickens originated in the USA, and the term as generally used today came into Britain with the broiler chicken in the 1950s.

CHICK

A chick is a newly hatched chicken. In the traditional period of the history of the industry the bird was called a *chick* from day old until its removal from the brooder, (Robinson, 1948), and this definition is still useful. During the traditional period the term *chicken* was occasionally used as a synonym for *chick*.

LAYER; LAYING HEN

This is a female chicken of a strain selected for egg production. It may be of either a traditional breed or a modern commercial hybrid. Although the term *hen* had an age specific meaning in the traditional period, (see the section on the definition of *pullet* below), it is now often used more loosely to mean any laying chicken during its laying life. The age at point of lag is usually recommneded by the breeding company, but is profoundly affected by lighting. It is approximately 16 to 18 weeks (see Chapter 8).

PULLET

In the traditional period this term usually meant a female laying type chicken up to the end of its first laying season, (Robinson, 1948), after which it became a *hen*. In the conventional period, when the majority of birds were kept for only one laying year, it generally came to mean a young bird of a laying strain before it reached sexual maturity, hence the term *pullet rearing*. However it is still sometimes used to include the early part of the laying stage of the bird's life, hence the occasional reference to small eggs from early lay as *pullet eggs*. For the purposes of this book we propose to limit the application of the term to the immature bird, during rearing, before the start of lay.

Robinson (1948) pointed out that fancy breeders had a different definition of a pullet, namely a female bird hatched on or after 1 November of the previous year.

TABLE CHICKEN

A table chicken is a breed or an individual chicken intended for meat production.

5. Food supply chain terminology

HAZARD ANALYSIS CRITICAL CONTROL POINT (HACCP)

This is a formally documented set of disciplines and procedures used to minimise risks, particularly risks to food safety. It involves the identification of points in the process which are especially vulnerable to hazards (the critical control points) and formalises the precautions to be taken in their management. There are specialists in the appropriate analyses.

QUALITY ASSURANCE AND TRACEABILITY

These two terms both relate to modern methods of offering consumers assurance that they are buying what they wish to buy. The concepts are particularly important for niche products. Consumers wish to know where the products came from, and to be reassured about the production methods, the standard of animal welfare and even the geography of their origins. Sometimes these attributes are referred to as the *provenance* of the foods. *Quality assurance* normally consists of a degree of control and documentation of the production practices. There are named schemes whose products carry logos and for which written standards and rules are enforced. *Traceability* describes the processes of recording the progress of the product along the food supply chain from producer to consumer, so that its origins can be traced.

PROTECTED DESIGNATION OF ORIGIN, PROTECTED GEOGRAPHICAL INDICATION AND TRADITIONAL SPECIALITY GUARANTEED

These are EC schemes, (under EC Regulations 2081/92 and 2082/92), for the protection of the names of local and regional speciality food products. They are operated by the Department for Environment, Food and Rural Affairs (DEFRA), and by the corresponding organisations for Scotland and Wales. For further information for England and Northern Ireland see:

DEFRA, Room 20, Whitehall Place East, London SW1A 2HH (020 7270 8061);

for Scotland see: Room 243, 47, Robb's Loan, Edinburgh EH14 1TW (0131 244 6387)

and for Wales see: Agriculture Policy, Division 2, Cathays Park, Cardiff (029 2082 3376).

6. Experimentation

EXPERIMENT

In the course of the evolution of these products there has often been a need to test ideas or techniques. In this book an *experiment* is taken to mean a formal test involving randomised and replicated *plots*, and usually designed to test a hypothesis. In commercial practice unreplicated testing of ideas or techniques often, and quite rightly, takes place, but this is a different kind of test since it does not quantify the element of chance in the results. A *plot* is a recording unit, and in the present context it is usually a pen or a group of birds.

References

Gordon, S.H., Charles, D.R. and Green, G. (2001) Metabolic age: a basis for comparison of traditional breeds of meat chickens. *British Poultry Science*, **42** (supplement): 118-119

Laird, A.K. (1966) Postnatal growth of birds and mammals. *Growth* **30**: 349-363

Lampkin, N. (1997) Organic poultry production. Report to MAFF, CSA 3699

Lawrence, T.L.J. and Fowler, V.R. (1997) *Growth of Farm Animals*. CAB International, Wallingford, UK

Robinson, L. (1948) *Modern Poultry Husbandry*. Crosby Lockwood, London, UK

Taylor, St C.S. (1965) A relationship between mature weight and time taken to mature in mammals. *Animal Production* **7**: 203-220

Taylor, St C.S. (1968) Time taken to mature in relation to mature weight for sexes, strains and species of domesticated mammals and birds. *Animal Production* **10**: 157-169

Taylor, St C.S. (1980) Live-weight growth from embryo to adult in domesticated animals. *Animal Production* **31**: 223-235

Wilson, B.J. (1977) Growth curves: their analysis and use. In: *Growth and Poultry Meat Production*. Edited by Boorman, K.N. and Wilson, B.J., Poultry Science Symposium No.12, British Poultry Science Ltd.

HISTORICAL BACKGROUND

Introduction

For centuries livestock farming was integrated with crop rotation and with whole farm systems. Poultry production has often been somewhat detached from all this, however, and in the conventional period (as defined above) it became even more separated. This chapter offers a brief account of the development of poultry production systems in the UK, and describes some of the historical background to some technical problems which have recently come to light during current attempts to reintegrate poultry with the land and to develop modern niche production.

Seven histories and descriptions of the development of the UK poultry industry have been published in the past 15 years, and this chapter draws heavily upon them. They were that of Hewson (1986), Telford *et al.* (1986), Charles (1996a and b), Charles and Tucker (1997), Whittle (1998) and Charles (2002). Historical, scientific and technical aspects of the development of livestock and poultry housing practices were also described by several authors edited by Wathes and Charles (1994).

The history of sustainable livestock systems

The maintenance of soil fertility, including by the integration of livestock with the land, has a long and interesting history.

Over the centuries from before the Norman conquest until enclosure, the attempt to sustain soil fertility was a cardinal principle of the agrarian economy and thus of the entire social structure of rural England.

Each village had three, or occasionally two, communal open fields in each of which the freemen held randomised strips, redolent of the plots of modern field experiments. One field was left fallow each year to permit it to recover from cereal cropping. Presumably its recovery was due to the action of factors such as leguminous weeds, resembling the way in which Cormack (1999) used a red clover mulch during conversion of a site to organic during recent tests. In some areas of the country there were occasional applications of materials such as marl, if they were available locally. In addition the free roaming livestock of the villages had access to the fallow and would have therefore deposited manure, so that mixed farming with livestock is a time honoured technology.

In some parts of the country, particularly in the English Midlands, the layout of the strips can still be seen in the ridge and furrow on what is now grassland. The enclosure movement was motivated by many things, among which was the desire of some of the villagers and land owners to exploit improved crop rotations (Prothero, 1936; Blaxter and Robinson, 1995). Enclosure consolidated farms and provided fields which were no longer divided into small strips. Because hedges were planted enclosure also improved the potential for the management of grazing animals within the rotations. Not all historical writers, at the time or since, have regarded enclosure as an unequivocally good thing, for sociological reasons, but such debates are not relevant to this book (see Charles, 2002, for more detail).

Significant events in the development of the maintenance of soil condition included the introduction of turnips and clover, and their consequences for livestock, in the 16th and 17th centuries. The exploitation of these crops was through sheep and cattle, whose manure contributed to soil fertility. Probably the best-known rotation was the Norfolk four course (wheat-turnips-barley-clover).

Livestock were crucial to improved rotations. Therefore livestock improvers, such as Robert Bakewell (1725-1795) were part of the same story (Pawson, 1957; Fraser, 1959; Stanley, 1995), but they were seldom interested in poultry.

The Board of Agriculture, established in 1793 was the forerunner of the Ministry of Agriculture, Fisheries and Food (MAFF). It staged Sir Humphrey Davy's lectures in 1803 on agricultural chemistry, this marking the beginning of a scientific approach to soil fertility.

Poultry in UK farming systems: early developments in poultry science

Poultry husbandry in the traditional period, (as defined above as the period from the end of World War I until the ending of the rationing of layers mash in 1953/4), used pasture and was sometimes integrated with crop rotations. For example at ADAS (then National Agricultural Advisory Service, NAAS) Gleadthorpe in the early 1950s laying hens were part of a six year rotation comprising potatoes-cereals-cereals-poultry-poultry-potatoes.

The authoritative technical writers of the period recommended that well managed good quality pasture could contribute to the nutrition of the birds (Robinson, 1948; Thompson, 1952). They therefore recommended pasture management practices conducive to grazing by the birds. This utilisation of pasture meant that the droppings were returned to the land.

If there was any single turning point time in the poultry industry's history it was probably 1954, when poultry mash came off the rationing imposed during and after World War II (Hewson, 1986). (Though note that Whittle, 1998, gave the date of the cessation of rationing as 1953). Thus from 1939 until 1953/4 most of the production was necessarily confined to small units whose size was limited by the feed supply.

The technology we now regard as conventional (see the definitions given above) began to develop after 1953/4, when consumers increasingly demanded more abundant and cheaper supplies as they emerged from deprivation, and when, in response, poultry businesses began to grow in both size and technical efficiency. Data quoted by Poultry World (1959) showed that in 1948 only 11% of the birds were in flocks of over 500, but by 1955 this had doubled to 22%. Data from MAFF (1994) showed that by 1988 70% of the layers were in flocks of over 20,000. By 1998 76% were in flocks of over 20,000, on 200 holdings. As the successful holdings increased in size some of the smaller and less successful dropped out, and thus the numbers of holdings with poultry has steadily declined during the period after 1953/4 (Table 1), yet in 1998 there were still 25,200 holdings with less than 5,000 layers, suggesting that there were many very small units (MAFF, 1999). The larger businesses tended to be specialised and not necessarily associated with very much land.

The severe constraints on poultry business activity imposed by World War II and its aftermath meant that little new development took place between 1939 and 1953/4. Thus in order to make observations about commercial versions of traditional systems we must examine the technologies of the inter-war years before 1939.

Table 1. NUMBERS OF HOLDINGS WITH POULTRY, THOUSANDS

1957	250[1]
1965	160[1]
1988	44[2]
1993	34[3]
1998	28[3]

1=Coles (1966)
2=MAFF (1994)
3=MAFF (1999)

Serious commercial production of eggs got under way after World War I, supported by local authorities, who provided both county laying trials and county poultry instructors; by the agricultural colleges and universities; and by trade bodies such as the Scientific Poultry Breeders Association, whose journal *Eggs* contained scientific material on subjects such as nutrition.

After World War II a MAFF funded advisory service (NAAS; later ADAS) offered technical and management advice, and the Agricultural Research Council and the Universities provided scientific support for the growing industry.

Most of the county agricultural colleges, and the national agricultural colleges such as Harper Adams, carried out research and development (R&D). Although the journal *British Poultry Science* was not launched until 1960, the American research and educational infra-structure was well developed before World War II, and the journal *Poultry Science*, first published in the early 1920s, was widely read in UK, and its messages applied in the British industry. The development and promulgation of what we have now decided to call traditional poultry technology must have been considered rather important.

After 1953/4 the dictates of both economics, and the need to greatly expand the national egg supply in order to meet growing demand, led to the rapid growth in the popularity of intensive indoor systems of egg production, as shown in Table 2. For market reasons there has recently been a slight reversal of this long-term trend, and the proportion of eggs produced on range has increased of late. Broiler production has been virtually 100% intensive since its introduction into Britain in 1956. Despite the fact that the free range proportion of broiler production is very small, though figures are hard to come by, there is now a great deal of interest in developing such niche markets. Free range turkey production is a well established market, but it is a small proportion of the total UK turkey production. By contrast it seems probable that parent and grand parent flocks of broiler strains will remain intensive, for reasons of hygiene and biosecurity as well as for the control of breeding programmes and the application of specialist management skills.

Table 2. CHANGES IN EGG PRODUCTION SYSTEMS, % OF EGGS

	Cages	Litter or barn	Free range
1951	8	12	80
1963	27	56	17
1980	95	4	1
1990	85	3	12
1993	85	3	12
1995	86	3	11
1999 (provisional)	79	5	16
2000 (provisional trade estimates)	71	6	23

Data from MAFF, from the Museum of the British Poultry Industry, Sacrewell, Peterborough, and from the trade.

Organic eggs are perhaps by now about 2 to 4% of the total egg supply. There are also organic table poultry in small but increasing quantities, but the proportions are uncertain.

The growth of intensivism separated poultry production from the land. The manure had to be disposed of, so it generally found its way onto land, even if it changed hands in the process, but it was merely a by-product commodity to be traded and used as well as possible, and subject to losses of plant nutrients during storage and transport. The birds were not usually regarded as integral parts of crop rotations or of the maintenance of soil fertility on arable and grassland holdings. For practical purposes poultry enterprises began to be seen as separate systems during the conventional period of the evolution of the industry.

A return to production methods reminiscent of those of the pre-war period undoubtedly offers potentially useful niche markets for the hard pressed UK poultry industry, and it is an appropriately demand led trend. As pointed out by Beaumont (1997) *producer push* has been replaced by *consumer pull* in modern food markets, and industry sectors ignore this at their peril. However the amount of technical endeavour described above suggests that the idea that the traditional period was a low-tech age is a modern myth.

On the contrary it had very high tech aspirations, spurred on by consumer needs for egg supplies. Hewson (1986) pointed out that even in 1938 nearly 30% of eggs were still imported.

Historical technical aspects of some key issues

The MAFF funded study group on organic pullet production, held at ADAS Gleadthorpe in 2000, identified a number of potential difficulties with the outdoor rearing of replacement pullets for egg production. Many of the difficulties would have seemed familiar to poultry keepers in the traditional period prior to 1939. Since they offer useful examples of historical aspects of technical and scientific issues arising out of the current interest in niche poultry production, they are quoted as follows.

1. BROODING

A MAFF funded project on extensive table chickens provided ADAS Gleadthorpe with the opportunity for experience with brooding in small low tech sheds, of a type popular until the 1950s but now staging a come back; and for a comparison with controlled environment brooding.

The provision of comfortable temperatures under the heater, as indicated by chick clustering behaviour (Charles, 1986), while providing draught free (<0.15 m/s) yet generous air change rates, proved to be very difficult, even in summer (though July 2000 was the coolest July for many years). This was because as weather conditions changed, in particular temperature, wind speed and wind direction, the heaters and air inlets needed constant adjustment. However staff cannot remain on site all night, so both bird losses and fuel consumption were higher than in controlled environment brooding, where thermostatically and automatically controlled ventilation and heating equipment was provided to meet the published estimates of the physiological needs of the chicks.

Note that the sheds were fitted with insulation to conserve heater fuel and, more importantly, to prevent condensation, and thus risks of damp unhealthy litter. This concession to conventional controlled environment technology was considered appropriate to the project and to niche markets since, although not a typical feature of the traditional low tech housing of, say, the 1930s, it had been introduced into the traditional industry, to prevent frost, by the time of publication of a MAFF booklet on poultry housing of as early as 1955.

Sources such as Hewson (1986) and Robinson (1948) show illustrations of a wide variety of brooders in use in the traditional period, though many resemble modern canopy brooders quite closely, apart from using paraffin, whereas propane is now the usual fuel. Electric brooders were popular. The chicks were often let out onto grass from a few days of age. It seems that by 1948 few were using broody hens except small-scale poultry keepers. Larger units used battery cage brooders with the chicks in heated drawers (Robinson, 1948). There were also tier brooders, with typically two tiers of cages with 100 chicks per tier, heated by oil, hot water or electricity, and there were cages with indoor wire floored runs. Cage systems were said to be relatively disease free but to be associated with poorer feathering.

It appears that during the traditional period a popular approach was to brood indoors and then to rear on range. This probably reflected concern about early mortality. Poultry World (1959), in a manual of husbandry, pointed out the importance of chilling as affecting mortality. A high death rate during the first few days of brooding was cited. Thus the manual noted that for over 20 years before the time of writing in 1959 the majority of chicks had been reared intensively for at least the first four weeks of life. Yet until well into the 1960s rearing on range after brooding remained popular, partly out of a belief in the ability of outdoor rearing to provide hardy pullets, and partly for reasons of capital cost. By about the late 1960s, however, commercial pullet readers switched to fully indoor methods, perhaps mainly in order to apply lighting programmes and feeding regimes more reliably, but also to reduce fossil fuel consumption.

In recent times there has been some confusion over the use of the term *deep litter*. During the traditional period deep litter meant litter built up and unchanged between batches of birds, and allowed to ferment and to compost within the house. This generated heat, which, according to Poultry World (1959), the chicks appreciated and which reduced chilling, but not surprisingly coccidiosis was a problem, with bloody droppings sometimes observed as young as six days. Drugs were then administered. Thus perhaps the idea that it was modern (*conventional*) systems which gave rise to the need for drugs may be another modern myth.

Needless to say it was soon preferred that indoor litter rearing systems used new litter for each batch, though it was no longer necessary for it to be particularly deep. Thus for describing indoor systems as used for most of the periods under review, and certainly since the advent of the broiler industry in 1956 (see Whittle, 1998, for a history of the introduction of broilers) the term *deep litter* is incorrect, and a better term might be simply *litter*.

Another serious problem with deep re-used litter was ammonia. Cotterill and Winter (1953) attributed outbreaks of ocular inflammation amongst broilers in USA at that time to ammonia from built up litter, and recommended the use of fresh litter for each batch. Saunders (1958) described an outbreak of keratoconjunctivitis in Huntingdonshire where there were no apparent vitamin A deficiencies nor infectious disease causes, and attributed the condition to ammonia. Charles and Payne (1964; 1966) warned of the dangers of ammonia to the birds and to their performance.

Brooding ISA 657 chickens in low tech free-range facilities at ADAS Gleadthorpe

2. REARING SYSTEMS AND DISEASE CONTROL

Hewson (1986) mentioned that in the 1930s two stage rearing, with fold unit brooders and runs moved daily, was the only way to control coccidiosis.

Brooding ISA 657 chickens in low tech free-range facilities at ADAS Gleadthorpe

The second stage was either a larger fold unit without heaters, or it was Sussex arks or range shelters. He noted that battery brooders kept the chicks and the workers in the dry for up to four weeks. In the post war period three stage rearing was introduced, with a battery brooder followed by a hay box fold and then a slatted floor verandah. Photographs of these were provided in Hewson's paper. It was later realised that moving was stressful to the birds, so indoor floor rearing systems were developed. Thompson (1952) published photos and diagrams of brooding systems, including the servicing of quite large numbers of outdoor fold brooders in commercial practice.

In a chapter on disease prevention Howes (1939) recommended the application of lime to the land used for poultry and the thorough sprayed disinfection of the houses with "recognised disinfectants", some of which could be "embodied in limewash".

Mortality appears to have been severe by modern standards. In a chapter on budgeting Robinson (1948) advised an assumption of 10% losses in the brooder house, followed by 10% during growing, plus 10% culls: a total of 30% mortality. It was stated that good rearers with healthy stock should lose less, but that in order to have the required number of layers 30% should be allowed for. This must have been a high estimate, however, since a report of the Middlesex County Laying Trials for 1936/7, by Worthington, reported the loss of 26 birds out of 114 (23%) and describes this as "deplorable, and a striking commentary on modern (*sic*) methods of rearing".

3. CONTINUITY OF SUPPLY

It may sometimes be assumed that the application of lighting patterns for egg production is part of conventional (post 1953/4) technology. The definitive understanding of the process could be said to date from the late 1950s (*i.e.* within the conventional period), with such landmark work as that of Morris and Fox (1958a and b) at Reading University, but the basic technology is much older than that, and well rooted in the traditional period.

It may be difficult for the last few generations of British consumers to believe that eggs used to be either unavailable or imported in the winter, yet surprisingly recently cookery books advised their readers that eggs were cheaper and easier to obtain from Easter to June (*e.g.* Laskie, 1950). Up to the 1950s large quantities were imported and preserved at home in pots of water-glass (sodium silicates) for winter use.

Whetham (1933) noticed that the seasonality of egg production followed daylength. He plotted daylengths and egg production for different latitudes, and realised that the operative factor was the seasonal change in daylength and not the daylength *per se*. Before that time the function of longer days was generally believed to be merely to give the birds more time to eat (e.g. Fairbanks and Rice, 1924; reviewed by Hammond, 1960).

Parkhurst and his colleagues at Harper Adams Agricultural College developed practical winter lighting programmes for layers in 1928 and Howes (1939) strongly recommended the application of artificial lighting for layers in the winter months. Interestingly although leaflets aimed at domestic poultry keepers, such as that of ADAS (1979), often mentioned the need for supplementary light in winter even for very small scale houses, war time booklets such as that by Powell-Owen (undated, *circa* 1941), aimed at the numerous home poultry keepers of the time, did not mention lighting, presumably because of the blackout regulations. (Incidentally this 30 page booklet was priced at 6d., or 2.5 p!).

References

ADAS (1979)(first published much earlier, date unknown, probably 1940s)
        Domestic poultry keeping.
Beaumont, J. (1997) The consumer has spoken. Keynote address. BOCM
        PAULS Conference, The Belfry, Wishaw, UK

Blaxter, K.L. and Robertson, N. (1995) *From dearth to plenty. The modern revolution on food production.* Cambridge University Press, Cambridge, UK

Charles, D.R. (1986) Temperature for broilers. *World's Poultry Science Journal* **42**: 249-258

Charles, D.R. (1996a) A brief history of the UK poultry industry. In: *UK Branch 50th Anniversary.* pp. 45-51 Edited by Fisher, C. and Hann, C.M., World's Poultry Science Association

Charles, D.R. (1996b) The Museum of the British poultry industry. *Journal of the Royal Agricultural Society of England* **157**: 197

Charles, D.R. (2002) *Food, farming and the countryside: past, present and future.* Nottingham University Press (in press).

Charles, D.R. and Payne, C.G. (1964) The effects of ammonia on the performance of laying hens. *World's Poultry Science Association 2nd European Poultry Congress,* Bologna, 109-112

Charles, D.R. and Payne, C.G. (1966) The influence of graded levels of atmospheric ammonia on chickens. 1. Effects on respiration and on the performance of broilers and replacement growing stock. *British Poultry Science* **7**: 177-187

Charles, D.R. and Tucker, S.A. (1997) The poultry industry in the United Kingdom. *Journal of the Royal Agricultural Society of England* **158**: 175-183

Coles, R. (1966) Size changes in laying flocks. *The poultry review.* May and Baker Ltd., Dagenham, UK

Cormack, W.F. (1999) Testing a stockless arable rotation on a fertile site. *Crop Rotations Design Workshop*, Borris, Denmark, 1-9

Cotterill, O.J. and Winter, A.R. (1953) Some nitrogen studies of built up floor litter. *Poultry Science* **32**: 365

Fairbanks, F.L. and Rice, J.E. (1924) Artificial illumination of poultry houses for winter egg production. *Cornell Extension Bulletin* 90 (Abstract)

Fraser, A. (1959) *Beef cattle husbandry.* Crosby Lockwood, London, UK

Hammond, J. (1960) (first published 1940) *Farm animals.* Edward Arnold, London, UK

Hawk, W. (1910) *Poultry keeping for profit.* Cornwall County Council

Hewson, P.F.S. (1986) Origin and development of the British poultry industry: the first hundred years. *British Poultry Science* **27**: 525-540

Howes, H. (1939) *Modern poultry management.* Macmillan, London, UK

Laskie, M.G. (1950) *Cookery for girls.* English Universities Press

Ministry of Agriculture, Fisheries and Food (1955) *Poultry housing*. Bulletin No. 56, HMSO

Ministry of Agriculture, Fisheries and Food (1994) *Agriculture in the United Kingdom 1993*. HMSO

Ministry of Agriculture, Fisheries and Food (1999) *Agriculture in the United Kingdom 1998*. The Stationery Office

Morris, T.R. and Fox, S. (1958a) Light and sexual maturity in the domestic fowl. *Nature* **181**: 1453-1454

Morris, T.R. and Fox, S. (1958b) Artificial light and sexual maturity in the fowl. *Nature* **181**: 1522-1523

Parkhurst, R.T. (1928) Artificial light for late hatched pullets. *Eggs*, Scientific Poultry Breeders Association, December 1928, 270-271

Pawson, H.C. (1957) *Robert Bakewell: pioneer livestock breeder*. Crosby Lockwood, London, UK

Poultry World (1959) *The poultry handbook*

Powell-Owen, W. (1941?) *Poultry keeping on small lines*. The Home Front series. Country Life Ltd.

Prothero, R.E. (1936) *English farming past and present*. 1961 edition enhanced by Fussell, G.E. and McGregor, O.R., Heinemann and Frank Cass Ltd., London, first published 1912

Robinson, L. (1948) *Modern poultry husbandry*. Crosby Lockwood, London, UK

Saunders, C.N. (1958) Keratoconjunctivitis in broiler birds. *Veterinary Record* **70**: 117

Stanley, P. (1995) *Robert Bakewell and the Longhorn breed of cattle*. Farming Press Books, Ipswich, UK

Telford, M.E., Holroyd, P.H. and Wells, R.G. (1986) *History of the National Institute of Poultry Husbandry*. Nuffield Press, Oxford, UK

Thompson, A. (1952) *The complete poultryman*. Faber and Faber, London, UK

Wathes, C.M. and Charles, D.R. (1994) *Livestock housing*. CAB International, Wallingford, UK

Whetham, E.O. (1933) Factors modifying egg production with special reference to seasonal changes. *Journal of Agricultural Science* **23**: 383-419

Whittle, T.E. (1998) *A triumph of science: a 70 year history of the UK poultry industry*. Poultry World

Worthington, J. (1937) *Report of the twelfth egg laying trial, 1936/37*. Middlesex County Council

2

MARKET BACKGROUND

Recent trends in consumer spending on food products

In developed countries food accounts for only a small proportion of the household budget, but this was not always so. In the UK as recently as the 1950s household food purchases accounted for about 28% of income (MAFF, 1997). In 1960 total expenditure on food, (i.e. household food plus food eaten outside the home), accounted for 31% of income (Office for National Statistics, (ONS), 1997).

When food was as expensive as that most consumers wanted it to be as cheap, nourishing and abundant as possible. But things are different now. By 1998/9 food accounted for only 17% of total household expenditure (ONS, 1999). Many consumers now want something more from their food, and this socio-economic development has been a significant driver of change in the market.

Producers and purveyors of foods, including poultry products, may miss opportunities if these social changes are ignored. It is now consumers rather than producers who drive markets. Furthermore change is a key word, and market conditions are far from stable. Van Trijp and Steenkamp (1998), quoting Hughes (1994), noted that over 25% of retail food sales in the USA consisted of products introduced in the previous five years. Product innovation and change are characteristics of the modern food market, including poultry products.

However markets are not the only engines of change in the development of aspects of food production. Lampkin (2001) listed five drivers of growth in organic agriculture: regulatory matters, market growth, policy support, insecurity in conventional agriculture and developments on grassland farms. Some of these may affect poultry products.

Classifications of food consumers

SOCIO-ECONOMIC CLASSIFICATIONS

Although on average modern consumers in developed countries are affluent, averages hide wide variations, and there are several well-known socio-economic classifications. Two examples may suffice.

1)   The ONS categorise household income groups by quintiles. In 1998/ 9, according to ONS (1999), the poorest quintile in the UK had an average gross weekly household income of £94 per week. A market for commodity poultry products is therefore likely to persist. By contrast the best off quintile had an average gross household income of £1057 per week. These better off consumers are unlikely to be content with commodity products, but they may be interested in niche poultry products.

2)   MAFF (2000), and many market researchers, classify consumers by the following socio-economic groups.

Table 1. SOCIO-ECONOMIC CLASSES, BY GROSS WEEKLY INCOME OF THE HEAD OF THE HOUSEHOLD, £

Class		*Weekly income, £*
A1		>980
A2		655-980
B		345-655
C		165-345
D		<165
Households without an earner,	E1	>165
	E2	<165
Old age pensioners		

Presumably the market for the higher priced premium niche poultry products will often be amongst classes A1, A2 and B.

OTHER CLASSIFICATIONS

Classifications by age, occupation and lifestyle are often quoted by marketing specialists. For example age groups could be classified by life stage as follows:

Class 1. Infants and children
Class 2. Teenagers
Class 3. Young singles
Class 4. Young families
Class 5. Pre-retirement post family
Class 6. Retired
Class 7. Elderly

It is not clear which of these may be most interested in niche poultry products.

Types of food purchase

EATING AND SHOPPING STYLE

Hughes and Ray (1999) recognised a distinction between purchases made in what they termed *drudge mode*, usually on a week day and in a hurry, and purchases in *leisure mode*, usually at the weekend. In leisure mode there is time to think about the food and its origins, and niche products may sometimes be purchased in this mode.

Other distinctions also influence purchases. A considerable proportion of the food spend, including on poultry products, is now on eating out. This includes a spectrum of eating styles and eating occasions, from fast food purchased cheaply, eaten quickly and often on the move, to leisurely dining out. Niche poultry products probably appeal mainly, but not exclusively, to the latter.

Food also has important, and ancient, social and cultural significances. Niche products are likely to appeal to purchasers shopping for a family occasion or a socially significant event. Gofton (1996) reviewed the social significance of foods.

PROVENANCE

Modern consumers are concerned about issues such as animal welfare, the environment, and healthy eating, including the rapidly growing interest in organic foods. There is considerable interest in the provenance of food, and a revival of regional and speciality foods. Developments such as farmers' markets, box schemes and internet shopping for speciality foods are recent manifestations of these interests. All of these may affect the market for niche poultry products. Organic foods have their own well defined infrastructure of certification bodies, some of whose logos are now well recognised by the food buying public, under the umbrella of the UK Register of Organic Food Standards (UKROFS).

ATTITUDES TO FOODS

Meulenberg and Viaene (1998) quoted a system devised by Grunert *et al.* (1996) classifying consumers by their attitudes to food, including uninvolved, careless, moderate, conservative, hedonistic and adventurous. It is not always clear which of these might require niche poultry products, but presumably the uninvolved will not. However involvement in issues such as animal welfare, the environment and organics may sometimes dominate purchasing behaviour.

PURCHASE MOTIVATIONS

Charles (1998) proposed a five class classification of purchase motivations, and later modified it to a six class classification (Charles, 2000). Table 2 is based on the six class classification.

Table 2. PURCHASE MOTIVATION CLASSES

Purchase type class	Purchase type description	Comments
1	Price motivated	Price sensitive staples and commodities
2a	Convenience and time saving - retail	Prepared foods, mixes *etc.*
2b	Convenience and time saving - eating out and fast foods	Sold by fast food outlets, cafes, roadside services, garage forecourts, pubs, institutional catering, sandwich bars, *etc.*
3	Healthy eating motivated	
4	Interest, variety, theme	Includes regional and speciality foods
5a	Leisure and pleasure - dining out	Restaurants, pubs
5b	Gifts and souvenirs	
6	Social and emotional	Includes family occasion meals

Niche poultry products are not Type 1, but may, perhaps sometimes be Types 2, 3, 4, 5a, 5b or 6. Note the difference between 2b and 5a, though both apply to eating outside the home.

Market sensitivity and market potential

None of this means that the market is unlimited, or that current producer premiums are necessarily stable or permanent. For example Lampkin (1997) reported that at that time the demand for organic poultry products exceeded supply, leading to premiums of 50% over standard free range prices. However he offered the important warning that significant expansion could lead to price volatility. Premium priced niche markets are, by definition, small. Lampkin (1999) illustrated the effects of price volatility on gross margins for organic eggs and chickens by means of sensitivity analyses. Profitability was shown to be very sensitive to fluctuations in product sale price, feed costs and feed conversion. Some detailed sensitivity analysis figures, and some example enterprise costings, are quoted in later chapters, but the point of marketing background significance is sensitivity.

Many of the scientific issues reviewed in this book have substantial potential effects on both technical efficiency and feasibility, and are therefore relevant to the potential viability of enterprises, and to addressing at least some of the input side aspects of the volatility of profitability. It is markets, however, which dominate the potential for profitability, and they are likely to continue to do so.

References

Charles, D.R. (1998) *Food purchasers - who are they and what do they want?* Middle England Fine Foods workshop paper.

Charles, D.R. (2000) *Egg purchasers - who are they and what do they want?* BOCM PAULS conference, Wakefield

Gofton, L. (1996) Bread to biotechnology: cultural aspects of food ethics. In: *Food ethics*. Edited by Mepham, B., Routledge, London, UK, pp. 120-137

Grunert, K.G., Baadsgaard, A., Larsen, H.H. and Madsen, T.K. (1996) *Market orentation in food and agriculture.* Kluwer Academic Publishers, Boston (Abstract)

Hughes, D. (1994) *Breaking with tradition: building partnerships and alliances in the food industry*. Wye College Press, UK

Hughes, D. and Ray, D. (1999) *Developments in the global food industry. A twenty first century view*. Wye College. University of London, UK

Lampkin, N. (1997) *Organic poultry production*. Report to MAFF, CSA 3699

Lampkin, N. and Measures, M. (1999) *Organic farm management handbook*. University of Wales and Elm Farm Research Centre

Lampkin, N. (2001) Future developments in organic farming - implications for the animal feed industry. In: *Recent advances in animal nutrition*. Edited by Garnsworthy, P.C. and Wiseman, J., Nottingham University Press, Nottingham, UK, pp. 151-159

Ministry of Agriculture, Fisheries and Food (1997) *National food survey 1996*. Stationery Office, London

Ministry of Agriculture, Fisheries and Food (2000) *National food survey 1999*. National Statistics. Stationery Office, London

Meulenberg, M.T.G. and Viaene, J. (1998) Changing food marketing systems in western countries. In: *Innovation of food production systems*. Edit. Jongen, W.M.F. and Meulenberg, M.T.G., Wageningen Pers, 5-36

Office for National Statistics (1997) *Family expenditure survey*. Stationery Office, London

Office for National Statistics (1999) *Family spending*. Stationery Office, London

van Trijp, J.C.M. and Steenkamp, J.E.M.B. (1998) Consumer oriented new product development: principles and practice. In: *Innovation of food production systems*. Edited by Jongen, W.M.F. and Meulenberg, M.T.G., Wageningen Pers, Wageningen, The Netherlands, pp. 37-66

3

PROTEIN SOURCES FOR ORGANIC AND NICHE POULTRY

Summary

In the EU and the UK poultry feeds are often composed of cereals as energy sources, together with soya as a protein source and a vitamin and mineral mix. There are aspirations to use home grown protein sources in poultry feed in place of imported soya, particularly in organic and sustainable systems.

Ultimately this will be achieved when there is an adequate and reliable supply of home grown protein material that is both suitable and consistent in quality. One of the initial steps is to identify those feed ingredients that have a protein content and amino acid complement such that when included in rations they permit economic optimum levels of growth or egg production. However, additional factors such as amino acid availability, metabolisable energy content, fibre content, digestibility and type and quantity of anti-nutritive factors (ANFs) will influence their maximum inclusion rate. Furthermore, the vitamin and mineral content of feedstuffs will need to be considered as there is a preference for natural supplies of these nutrients in organic poultry rations.

Natural means of enhancing immunocompetence in birds on range will be important as will product quality assurance. Thus a feed ingredient that has a low protein content or a deficit of one essential amino acid may be considered valuable if it has other useful attributes.

The literature on vegetable protein sources grown in the UK and in the EU has been reviewed in this chapter. Several materials appear to offer potential as feed ingredients for organic poultry rations, including peas, beans, lupins, linseed, rapeseed and naked oats. However there are ANFs in most of them so considerable care is needed with their use. Tables have been prepared of nutrient analysis, ANFs and suggested maximum dietary inclusion levels. Some sources lend themselves to being grown organically.

Peas seem to be the most promising potential feed protein source, particularly if micronised and supplemented with lysine. They can probably be incorporated in broiler diets at up to 250 to 300 g/kg and in layers diets at up to 150 to 200 g/kg.

Some reports suggest that modest levels of sweet lupins (200 g/kg) might also replace soya in layers feeds. However, due to wide variation between lupin cultivars and in the treatment of raw materials, and therefore in their nutrient analysis, it is not possible to provide definitive universal recommendations. Thus the maximum inclusion rates suggested for lupins depend on the quality of the ingredient used.

Beans do not appear to be a good alternative protein source. This is because of their low concentration of sulphur amino acids and the presence of ANFs.

Rapeseed meal produced from the double low (00) glucosinolate species may only be incorporated in poultry rations at small concentrations (no more than 100 g/kg in feeds for laying hens and 50 g/kg in broiler starter rations, possibly increasing to 80 g/kg in broiler finisher rations). This is because of egg taint problems and off flavours in poultry meat caused by the presence of ANFs in rapeseed.

In the future naked oat may be a reasonable alternative protein source. This would be dependent on establishing its maximum inclusion rate as a replacement in wheat-soya based diets as used in the UK.

Since peas appear to be potentially valuable their attributes and characteristics have been reviewed in more detail than those of some of the other ingredients.

This chapter does not consider the agronomic feasibility of growing crops in an organic farming system.

Introduction

For many years commercial poultry feeding practice in the EU and the UK has often been based on cereals for the provision of metabolisable energy (ME) and soya as a protein source, together with vitamin and mineral mixes. However there are aspirations to develop and maintain sustainable agriculture within the EU and the UK. One necessary step in achieving this is to become more self-sufficient, at farm level and/or at local level, in the production of plant protein sources for use in animal feeds.

The protein requirements of the EU 12 for animal feeding are considerable, at 18.7 Mt of protein equivalent (UNIP-ITCF, 1995). The Community deficit in protein rich raw materials, which was 88% in 1973, still remains high.

However, the substitution of soya may become relevant for reasons other than sustainability. It is expected that there will eventually be an increased requirement for soya for human consumption. This is based on a predicted increase in the world population from an estimated 6 billion in 1999 to 9.4 billion in 2050 (UN, 1998). There are forecasts of decreasing soya yields in North America due to global warming (Mannion, 1995). Thus, the combination of an increased demand for soya for human consumption and lower yields by the world's primary soya producer may reduce its availability for use in animal feeds.

Another important driving force is the lack of consumer confidence in GM crops. This is a particularly important issue in organic production. Segregation of a major commodity crop such as soya into traditional and GM beans requires separate production and handling facilities at every stage of the supply chain. It also requires traceability.

The possibilities for substituting soya in poultry feeds are good. Although soya has a high protein content, and this protein is well balanced except for a shortage of methionine (Larbier and Leclercq, 1992, translated into English and edited by Wiseman, 1994), the presence of ANFs requires the bean to be heat treated. ANFs disturb digestive processes, including reducing the utilisation efficiency of protein, carbohydrate and minerals and inactivating dietary vitamins (Huisman and Tolman, 1992).

When reviewing ANFs in plant proteins Huisman and Tolman (*loc. sit.*) suggested that most ANFs give the plant natural protection against attacks by moulds, bacteria and birds. Tabulating the relative presence of ANFs in various seeds allowed the authors to demonstrate the importance of certain ANFs according to seed type. Soya was found to contain medium to high levels of the following ANFs, the effects of which are given in brackets; trypsin inhibitors (reduced activity of trypsin, increased secretion of pancreatic enzymes), lectins (gut wall damage, immune response, increased loss of endogenous protein), antigenic proteins (interference with gut wall integrity, immune response) and alkaloids (neural disturbances, reduced palatability).

By comparison white flowered coloured peas do not contain tannins and Spring varieties sown early in the year have relatively low quantities of trypsin inhibitor (UNIP-ITCF, 1995). Helsper (1998) described the efforts of plant breeders to reduce the levels of tannins and trypsin inhibitors in legume seeds. The lysine content of peas is acceptable and their dietary energy value is compatible with the majority of diets employed in poultry production (UNIP-ITCF, 1995).

Sunflower seeds contain low levels of trypsin inhibitors and low to medium levels of phenolic compounds. Sunflower seed meal is a good protein source. Although it is deficient in lysine, it is very rich in sulphur amino acids (Larbier and Leclercq, 1992; Leeson and Summers, 1997). Its inclusion in conventional broiler rations is not limited by the presence of ANFs but by its moderate dietary energy value. This may be a virtue in a system where a slow growth rate is required, e.g. organic table chickens.

The options for substituting soya with home-grown protein sources in feeds for table birds and laying hens are discussed below. Recommendations on maximum inclusion rates are provided, although these mainly relate to varieties of relevant crops grown in a conventional, rather than an organic, farming system.

As with any raw materials there are factors which constrain their level of inclusion within a ration. These include the following: the crude protein content, amino acid composition and availability, metabolisable energy content, fibre content, digestibility and type and quantity of ANFs. The vitamin and mineral contents of potential ingredients are important as there is a preference for natural sources of these nutrients in feeds for organic livestock.

Wiryawan and Dingle (1995) proposed screening tests for samples of grain legumes. They included a 14 day growth bioassay and a modified limiting amino acid score. The latter was based on the ratios between mg amino acid/g protein in the legume to mg of the same amino acid/g dietary protein recommended by the National Research Council (1994).

Alternative protein sources

PEAS (*Pisum sativum*)

Peas are an increasingly important raw material in some European countries, in particular in France where they represent 11% of all raw materials used by the feed industry (UNIP-ITCF, 1995). The "protein crop" regulation introduced in 1978 stimulated an increase in dry pea production in the EU12 during the 1980s. Total production rapidly increased from less than 400 Mt in the early 1980s to approximately 4 500 Mt in 1994. Of the EU12 countries France is the most important in terms of pea production, accounting for approximately 79% of total production in 1994. In the same year, Denmark accounted for 8.5% of total production, England accounted for 6.2% of total production and Germany accounted for 3.5% of total production. The area used for dry pea production was 661 000 ha in France, 104 000 ha in Denmark, 80 000 ha in England and 45 000 ha in Germany. The total area used for dry pea production in the EU12 was 981 000 ha. This is a small area compared with that down to cereals. In 1998 3.43 million ha of cereals were grown in the UK alone (MAFF, 1999).

The area of land used for organic pea production in the UK in 1997 was reported to be only 20 ha (Soil Association, 1998).

Grosjean (1985) comprehensively reviewed work up to that time on the suitability of peas for animal feeding. He considered them to be satisfactory in poultry diets provided that methionine was added. Recommended maximum inclusion rates were found to vary between authors reviewed because of variation in the types of peas used and the ways in which the peas were treated. Values from 200 to 500 g/kg were quoted for broiler feeds and 200 to 300 g/kg for layers feeds. Ten years later UNIP-ITCF (1995) produced a major review on peas in animal feeding, with 144 references and much of the following comments on peas draws strongly on it.

The majority of dry peas destined for use in animal feeding are round-seeded, free from tannins and with low trypsin inhibitor activity (TIA). White flower coloured peas do not contain tannins unlike the coloured flowered varieties. Spring varieties sown between January and April are favoured because of their lower TIA and fibre contents than winter sown varieties.

The crude protein content of peas is variable. Between 1987 and 1994 the crude protein content was 241 ± 12 g/kg DM with the maximum range in one year approaching 70g/kg DM (UNIP-ITCF, 1995). The causes of variability are genetic, cultural and environmental in origin. However, the crude protein content of a variety may be established and reproduced when adopting similar cultural and environmental conditions. UNIP-ITCF (1995) cited work by Carrourée and Duchene (1993), which identified techniques and climatic conditions having a large effect on crude protein content. They found that early drought at the beginning of flowering followed by a period of high moisture generally led to low protein contents. Pea weevil attack or soil compaction below the seed bed, which reduce nitrogen fixation, are likely to lead to lower protein contents. Very early sowing dates will generally lead to higher protein concentrations.

The apparent nitrogen digestibility of raw peas in the adult fowl is estimated as being 75-80% (UNIP-ITCF, 1995 citing the work of Carré *et al.,* 1991). The true digestibility of nitrogen is 88% according to RPAN (1993). Pea proteins are predominantly water soluble (UNIP-ITCF, 1995). The water soluble proteins are principally globulins (approximately 60% of the total protein content) and albumins (approximately 25% of the total protein content). Ratios of globulins to albumins may vary and the variability is primarily of genetic origin rather than environmental (Baniel *et al.,* 1992). Both globulins and albumins are fairly rich in lysine. As a consequence the ratio of lysine:crude protein content is high. However, the crude protein and lysine contents of peas are less than that of soya bean meal but higher than that of cereals. Lysine is the first limiting amino acid in growth and egg production. Peas have low sulphur amino acid contents and a low tryptophan content compared with soya bean meal.

The amino acid composition of peas varies linearly with crude protein content according to work reviewed by UNIP-ITCF (1995). It has been claimed that it is possible to calculate the amino composition of peas according to the determined crude protein content using a function and constants published by Larbier and Leclercq (1992) or UNIP-ITCF (1995), quoting Mossé (1990). Table 1 shows their suggested linear relationships and also those reviewed by Grosjean (1985), who was quoting Baudet (1983). By comparison Igbasan *et al.* (1997) found that out of 10 essential amino acids, only for arginine was the content positively correlated with the crude protein content. Although the amino acid contents in Table 2 are inconsistent

Table 1. PUBLISHED COEFFICIENTS FOR PREDICTING THE AMINO ACID CONTENT OF PEAS BASED ON THE DETERMINATION OF CRUDE PROTEIN CONTENT

	a^1	a^2	a^3	b^1	b^2	b^3	r^1
Lysine	0.0598	0.0595	0.05952	3.58	0.364	0.364	0.992
Methionine	0.0075	0.0069	0.00691	0.65	0.078	0.078	0.935
Cystine	0.0059		0.00575	2.2		0.222	0.753
Threonine	0.0264	0.0273	0.02733	2.97	0.277	0.277	0.978
Tryptophan	0.0077	0.0071	0.00708	0.1	0.024	0.024	0.913
Leucine	0.0672	0.0670		1.08	0.109		0.99
Isoleucine	0.0374	0.0362		1.77	0.203		0.966
Valine	0.0424	0.0425		1.81	0.176		0.977
Histidine	0.0243	0.0247		0	-0.006		0.986
Arginine	0.1555	0.1538		-14.97	-1.455		0.972
Phenylalanine	0.037			2.85			0.975
Tyrosine	0.0226			2.89			0.939

1 = UNIP-ITCF quoting Mossé (1990)
2 = Larbier and Leclercq (1992)
3 = Grosjean (1985) quoting Baudet (1983)

Where, amino acid content, g/kg dry matter = a x crude protein (g/kg dry matter) + b

between authors they are very consistent when expressed as fractions of the lysine content, i.e.,

$$\text{methionine} + \text{cystine} = \text{lysine} \times 0.35 +/- 0.025 \qquad (1)$$

and

$$\text{tryptophan} = \text{lysine} \times 0.11 +/- 0.02 \qquad (2)$$

Digestibility of pea proteins is improved by pelleting (Carré *et al.,* 1987 and Conan *et al.,* 1992). Autoclaving has less effect on the digestibility of crude protein (Conan and Carré, 1989). Extrusion and fine grinding have very little effect according to Lacassagne (1988) and Conan *et al.,* (1992) respectively.

The dietary energy value of peas is compatible with the majority of diets employed in conventional poultry production. UNIP-ITCF (1995) published

Table 2. NUTRIENT ANALYSES OF ALTERNATIVE PROTEIN SOURCES (DM = DRY MATTER BASIS, AF = AS FED BASIS, NS = BASIS NOT STATED)

		Protein concentration g/kg	*ME, MJ/kg*	*Methionine + cystine, g/kg*	*Lysine, g/kg*	*Trypto- phan, g/kg*	*References*
Soya							
(full fat)	(NS)	375	16.2	11	24	5	Leeson and Summers (1997)
	(AF)	372		*2.9	*6.3	*1.3	van Kempen and Jansman (1994)
Peas	(NS)	235	10.7	5	16	2	Leeson and Summers (1997)
	(AF)	206		5.2	15.4	1.7	UNIP-ITCF (1995)
	(AF)	238	10.8	5.7	16.8	1.8	NRC (1994)
	(DM)	236		6.4	16.9	2.1	Larbier and Leclercq (1992)
	(AF)			5.5	15.2	0.9	McDonald *et al.*, (1995)
	(DM)	156-235		*2.7	*7.3	*0.9	Castell *et al.*, (1996)
Lupins	(NS)	345	12.6	9	17	4	Leeson and Summers (1997)
	(AF)	351		*2.2	*4.7	*0.8	van Kempen and Jansman (1994)
	(DM)	400		9.2	19.3	3.2	Larbier and Leclercq (1992)
	(AF)			9.2	17.0	1.8	McDonald *et al.*, (1995)
Beans	(DM)	291		5.8	18.3	2.3	Larbier and Leclercq (1992)
	(AF)			5.7	15.8	1.6	McDonald *et al.*, (1995)
Sunflower							
	(NS)	468	9.2	15	16	9	Leeson and Summers (1997)
	(AF)	167		*4.1	*3.5	*1.3	van Kempen and Jansman (1994)
	(AF)	233	6.5	10.0	10.0	4.5	NRC (1994)

Table 2. Contd.

	Protein concentration g/kg	ME, MJ/kg	Methionine + cystine, g/kg	Lysine, g/kg	Trypto-phan, g/kg	References
Sunflowers (contd.)						
(DM)	335		14.3	12.2	4.3	Larbier and Leclercq (1992)
(AF)			12.2	10.1	1.4	McDonald et al., (1995)
Rapeseed (canola) (NS)	375	8.4	13	22	5	Leeson and Summers (1997)
(AF)	196		*4.5	*5.4	*1.2	van Kempen and Jansman (1994)
(AF)	348	8.4	15.8	19.4	4.4	NRC (1994)
(DM)	386		18.1	22.0	4.6	Larbier and Leclercq (1992)
(AF)			15.5	21.5	1.7	McDonald et al., (1992)
Linseed (AF)	222		*3.6	*3.9	*1.4	van Kempen and Jansman (1994)
Naked oats						
(NS)	172	10.5 - 10.9	10.3	6.8	1.7	Maurice et al., (1985)

*g/16gN

an average metabolisable energy value for feed peas of 12.1 MJ/kg DM, which was determined using diets fed as meal to adult cockerels. They noted that the value of pea metabolisable energy in diets after pelleting was higher, being approximately 13.0 MJ/kg DM in the case of steam pelleting. This was said to be due to an improvement in the digestibility of starch from approximately 75-90% in raw peas to as high as 95% following heat treatment (Larbier and Leclercq, 1992). Starch is the predominant component of peas. Samples taken between 1987 and 1994 identified a concentration of 514 ± 15 g/kg DM (UNIP-ITCF, 1995). However, even after heat treatment pea starch is less digestible than cereal starch (Larbier and Leclercq, 1992). Digestibility of fibre is low (Longstaff and McNab,

1989), hence the improvement in metabolisable energy value of peas following dehulling. Autoclaving improved the metabolisable energy value of a pea variety with low meal metabolisable value by 13.5% (Conran and Carré, 1989). Lacassagne (1988) reported that extrusion had little effect on the metabolisable energy value of peas. Grinding through a screen size of 0.8 mm increased the metabolisable energy value of peas by 9.6% compared with a screen size of 4 mm (Conan *et al.*, 1992).

The metabolisable energy value of peas is lower in young chicks than in adult cockerels. The greatest difference is seen in meal diets, the reduction being 8.2% compared with only 0.5% in pelleted diets (Barrier-Guillot *et al.*, 1995). Askbrant (1988) reported that the metabolisable energy value of peas in laying hens fed meal was 7.4% lower than in adult cockerels. Work reported by UNIP-ITCF (1995) identified reductions between 2% and 5% in the metabolisable energy value of diets containing 18% peas in the laying hen.

Peas contain small amounts of oligosaccharides, these being between 63-75 g/kg DM according to UNIP-ITCF (1995). The major example is sucrose, representing between 30 and 40% of total soluble carbohydrates in peas, but there are also a-galactosides.

Igbasan and Guenter (1996a) carried out a detailed analysis of the effects of dehulling and micronisation on three pea cultivars. Literature they reviewed mentioned that potential negative effects on growth could be caused by heat labile protease inhibitors, tannins and lectins. For broilers fed 400 g/kg peas they found that birds fed micronised peas grew faster than those fed unmicronised peas. The addition of lysine improved performance further, probably because micronisation was found to have reduced the lysine content by 2.7%. For layers Igbasan and Guenter (1997a) found that 600 g/kg of untreated peas depressed egg production, but micronised peas did not. Igbasan and Guenter (1996b) found that pea chip by-products at 300 g/kg were unable to sustain broiler performance despite methionine supplementation.

The oil content of feed peas is low, being less than 2% (UNIP-ITCF, 1995). Triglycerides represent 90% of total lipids. Grosjean (1995) reported a predominance of linoleic acid. This fatty acid accounted for 50.2% of the total lipid content (UNIP-ITCF, 1995). Oleic, palmitic and linolenic acid are the other significant fatty acids, accounting for 20.0%, 12.3% and 12.1% respectively of the total lipid content.

Mineral compositions of raw materials were published by Larbier and Leclercq (1992). The calcium concentration of peas is lower than that of soya, the values being 0.9/kg DM and 3.1 g/kg DM, respectively. The total phosphorus concentration of peas is 5.2 g/kg DM and the digestible phosphorus concentration is 1.6 g/kg DM. Soya bean meal 50 has a higher total phosphorus concentration than peas, namely 7.8 g/kg DM but the digestible phosphorus concentration is marginally lower at 1.1 g/kg DM. The magnesium concentration of peas is lower than soya bean meal 50, at 1.4 g/kg DM compared with 3.2 g/kg DM. Calcium, phosphorus and magnesium are important minerals in maintaining egg shell quality and bone growth and development (Charles *et al.*, 1996).

Conventional diet formulation generally relies on the mineral and vitamin requirements of the bird being met by supplementing rations with premixes containing inorganic sources of minerals and synthetic vitamins. However, organic rations should aim to meet the birds' mineral and vitamin requirements using natural sources. Peas are rich in B group vitamins (UNIP-ITCF, 1995). Table 3 shows the vitamin content of peas compared with those of two other key ingredients.

Table 3. EXAMPLE VITAMIN CONTENTS OF FULL FAT SOYA, WHEAT AND PEAS, FOR COMPARISON

	Full fat soya	*Wheat*	*Peas*
Vitamin E (mg/kg)	40	13	1
Pantothenic acid (mg/kg)	10	12	11
Niacin (mg/kg)	15	55	37
Choline equivalent (mg/kg)	2174	2200	668
Riboflavin (B_2) (mg/kg)	2.4	1.1	2.2
Thiamin (B_1) (mg/kg)	1.8	4.8	7.4
Biotin (mg/kg)	0.31	0.11	0.16
Folic acid (mg/kg)	0.53	0.40	0.30
Vitamin A activity (10^3 IU/kg)	3.2		

(Based on Leeson and Summers, 1997)

Castanon and Perez-Lanzac (1990) used peas included at a concentration of up to 300 g/kg as a replacement for soya in diets for leghorn laying hens. Diets were equalised for energy and nitrogen concentration. They summarised their results algebraically as follows.

Feed intake (g/bird.day) = 111 + 0.01P   (r = 0.11)
Rate of lay (%) = 0.92 + 0.00002P   (r = 0.06)
Egg weight (g/egg) = 58.6 + 0.01P   (r = 0.36)
Feed conversion ratio = 2.05 - 0.0001P   (r = 0.15)

Where P = inclusion rate of peas, g/kg

The effect of increasing the concentration of peas in the diet on egg production, egg weight and feed conversion were all in favourable directions. However, the proportions of the variation accounted for by the equations (as indicated by r) were not large.

Table 4 summarises work on the presence of ANFs in several feedstuffs including peas and Table 5 summarises some of the key recommendations in the literature on inclusion levels.

LUPINS (*Lupinus* spp.)

Interest in the potential of lupins as protein sources for non-ruminants goes back to the early 1970s. Possibly the earliest reference was that of Gladstones (1970), who suggested that substantial substitution by lupins could be achieved without loss of production. Dun and Hopkins (1977) tested the inclusion of up to 150 g/kg sweet lupins (*L. angustifolius*), supplemented with methionine, in layers feeds at ADAS Gleadthorpe. Feed intake and egg output were reduced during early lay, and mortality increased progressively with lupin inclusion rate. The lupins obtained at the time contained less crude protein, methionine and lysine than expected.

More recently van Kempen and Jansman (1994) have reviewed work on lupins as an animal feedstuffs in detail. Potential problems which they pointed out included oligosaccharides of the raffinose family, a-galactosides, high levels of manganese in *L.albus* and pectins and alkaloids including lupanine, spartine, lupinine, and angustifoline. Whilst sweet varieties were

Table 4. ANTINUTRITIONAL FACTORS IN ALTERNATIVE PROTEIN SOURCES

	Antinutritional factors	*Possible palliatives*	*References*
Soya	Protease inhibitors, lectins, antigenic proteins, alkaloids	Heat treatment	van Kempen and Jansman (1994) Huisman and Tolman (1992)
Peas	Protease inhibitors, lectins, pectins, phenolics, tannins, haemagglutinins	Micronisation	Igbasan and Guenter (1996) Igbasan and Guenter (1997) Huisman and Tolman (1992) Castell *et al.*, (1996)
Lupins	Alkaloids, raffinose oligosaccharides, pectins, α-galactosides	Sweet lupin spp.	Olver and Jonker (1997) van Kempen and Jansman (1994) Huisman and Tolman (1992)
Beans	Tannins, trypsin inhibitors, lectins, phenolics, vicine/convicine	Heat treatment	Castanon and Perez-Lanzac (1990) Huisman and Tolman (1992)
Sunflower	Trypsin inhibitors, phenolics, phytates, chlorogenic acid, quinic acid		Huisman and Tolman (1992) van Kempen and Jansman (1994)
Rapeseed	Glucosinolates, phenolics, tannins, phytic acid, pectins, sinapines, euric acid	Low glucosinolate varieties	Khattak *et al.*, (1996) van Kempen and Jansman (1994) Huisman and Tolman (1992) McDonald *et al.*, (1995)
Linseed	Vitamin B_6 antagonist, trypsin inhibitor, cyanogenic glucoside		van Kempen and Jansman (1994)

Table 5. SOME PUBLISHED SUGGESTED MAXIMUM INCLUSION RATES OF VARIOUS VEGETABLE PROTEIN SOURCES (g/kg)

	Broiler feeds	Layers feeds	References
Peas	250-300	150-200, 300 for better egg taste	UNIP-ITCF (1995)
	To 4 weeks 50, after 4 weeks 100	100	Leeson and Summers (1997)
		100	Larbier and Leclercq (1992)
		300	Castanon and Perez-Lanzac (1990)
		200	Igbasan and Guenter (1997b)
Lupins	50	100	McDonald et al., (1995)
	To 4 weeks 80, after 4 weeks 100	150	Leeson and Summers (1997)
Beans		200	Castanon and Perez-Lanzac (1990)
	300	100	Larbier and Leclercq (1992)
			Jansman et al., (1993)
Sunflower	To 4 weeks 80, after 4 weeks 100	100	McDonald et al., (1995)
		150	Leeson and Summers (1997)
Rapeseed	50		McDonald et al., (1995)
	100 (double 00 varieties)	100 (double 00 varieties; white layers only)	van Kempen and Jansman (1994)
	Starter 50, finisher 80		Leeson and Summers (1997)

reported to be very low in alkaloids there is a risk of contamination by seeds of bitter varieties.

Olver and Jonker (1997) found that bitter lupins depressed the growth rate of broilers but sweet lupins were acceptable up to 400 g/kg inclusion rate. Van Kempen and Jansman (1994) reviewed work suggesting that satisfactory performance of layers could be obtained using up to 300 g/kg, and of broilers up to 400 g/kg of sweet lupins, though Naveed *et al.,* (1998) found that 400 g/kg of *L.albus* depressed broiler growth rate.

Brenes *et al.,* (1993), also working with *L.albus*, found that the addition of a protease enzyme, combined with a carbohydrase and an α-galactosidase, improved broiler weight gain. Dehulling improved broiler performance. Naveed *et al.,* (1998) found that the enzymes xylanase and cellulase prevented some of the depression in broiler weight gain which they had found at 400 g/kg, but the addition of a proteinase to the enzyme mixture was not beneficial, perhaps because it degraded the carbohydrases. Ferraz de Oliveira and Acamovic (1999) found that when feeding 400 g/kg of *L. angustifolius,* enzyme treatment reduced the amount of endogenous amino acid secretions. Gilbert *et al.,* (1999) supplemented 200 and 400 g/kg of the new *L. albus* varieties *Lucyane* and *Detn 12* with xylanase and cellulase, but unfortunately the growth rates of all treatments were low. Birds on treatments involving feeding lupin treatments experienced high mortality and the authors noted kidney problems. Using *L. albus* and *L. luteus* seeds Ferraz de Oliveira and Acamovic (1996) showed that incubation of the seeds with carbohydrases increased the dry matter digestibility.

Castanon and Perez-Lanzac (1990) quantitatively described their results with *L.albus* fed to laying hens at a dietary concentration of up to 200 g/kg, as follows:

Feed intake (g/bird.day) $= 113 + 0.05L$ $(r = 0.44)$
Rate of lay (%) $= 0.95 - 0.00004L$ $(r = 0.05)$
Egg weight (g/egg) $= 57.6 + 0.01L$ $(r = 0.25)$

where L = inclusion rate of lupins, g/kg.

It would appear that they found a very small depressing effect of increasing dietary lupin concentration on rate of lay but an increase in egg weight.

Combined with an increase in feed intake this resulted in a small increase in feed conversion ratio, as given below:

Feed conversion ratio = 2.07 + 0.0007L (r = 0.35)

The values of r show that the equations accounted for only small proportions of the variation. However, based on both their own results and earlier work published in Spain, they concluded that only peas and lupins are able to compete with soya bean meal as protein sources in poultry feeds. The literature reviewed here suggests that there is a degree of uncertainty in terms of performance when feeding lupins to poultry. Thus caution is recommended pending more work on practical diets under UK conditions, using varieties of *L. albus* that are likely to be available on the UK market.

BEANS (*Vicia faba*)

Huisman and Tolman (1992) reviewed the literature on the presence of ANFs in beans, and a summary can be found in Table 4. Tannins have often been of particular concern. Thus recent work has been aimed at examining the effect of tannin-binding agents on digestibility (Lamb and Acamovic, 1998). The low concentrations of methionine and linoleic acid in beans have also limited their utilisation historically.

The potential use of field beans in poultry rations was studied at ADAS Gleadthorpe during the 1970s. Particular attention was paid to problems associated with their low levels of methionine and of linoleic acid. Dun (1973, 1974 and 1975) found a progressive reduction in egg size and an increase in mortality when feeding beans at concentrations of 100 and 200 g/kg in layers feeds. In some experiments rate of lay was also depressed. Supplementation with methionine or linoleic acid did not prevent the detrimental effects on egg production. It is important to remember, however, that micronisation was not used, and since that time bean varieties have probably improved.

The latter may account for the high maximum inclusion rate suggested for growing chicks by Jansen *et al.,* (1993). The author proposed that beans of coloured flowered varieties may be included at up 300 g/kg, provided that the diets are nutritionally well balanced.

Castanon and Perez-Lanzac (1990) tested field beans as a replacement for soya bean meal in rations for layers. The diets were equalised for energy and nitrogen concentrations. They summarised their results algebraically as follows:

Feed usage (g/bird.day) = 119 - 0.04B   (r=0.66)
Rate of lay (%) = 0.98 - 0.00069B   (r = 0.74)
Egg weight (g/egg) = 58.5 -0.01B   (r = 0.71)
Feed conversion efficiency = 1.96 + 0.0003B   (r = 0.53)

where B = inclusion rate of beans, g/kg.

The values of r are higher for beans than those for peas. The regressions suggest that feed intake, egg production and egg weight all fell as the inclusion rate increased. Presumably some of the effects could have reflected the differences in intakes of essential amino acids which must have occurred between treatments.

A summary of published maximum inclusion rates for broilers and laying hens is provided in Table 5.

SUNFLOWER (*Helianthus annuus*)

Sunflower seedmeal is a good source of protein. Although it is deficient in lysine it is very rich in sulphur amino acids (Larbier and Leclercq, 1992; Leeson and Summers, 1997; van Kempen and Jansman, 1994). The inclusion rate of sunflower seedmeal in conventional broiler diets is probably limited less by ANFs than by its moderate dietary energy value. However, the development of de-hulled sunflower seedmeal has improved its nutritional value in poultry production. Typically there is an additional 0.84 MJ/kg of energy and 4% of crude protein when compared with standard sunflower seedmeal (Larbier and Leclercq *loc. sit.*). ANFs mentioned by van Kempen and Jansman (1994) were phytate, phenolic acids, chlorogenic acid and quinic acid.

Published maximum inclusion rates are provided in Table 5.

RAPESEED (*Brassica napus, B campestris*)

There has long been interest in the potential for rapeseed as a protein source. Overfield (1975), working at ADAS Gleadthorpe, investigated problems of egg taint occurring in the UK poultry industry at the time. He found that using six samples of rapeseed at inclusion rates of 100 g/kg, an average of 7% of the eggs were faintly tainted, as determined by a sniff panel. There were differences between samples of rapeseed and only certain stocks of birds were affected. Brown egg strains were more susceptible than white and one brown egg stock had 17% tainted eggs.

Trimethylamine (TMA) had previously been identified as the chemical present within egg yolks responsible for causing a fishy taint (Hobson-Frohock *et al.*, 1973). Butler and Fenwick (1984) in a review of TMA and fishy taint in eggs illustrated the dietary and genetic factors involved in its causation. Sinapine present within rapeseed meal is converted into choline and this is then converted by enteric bacteria in the large intestine of the laying hen into trimethylamine. Normally the TMA which is absorbed from the gastrointestinal tract of the fowl is rapidly converted to the odourless oxide by a microsomal enzyme (TMA oxidase) in the liver and kidneys and is excreted in this form (Lee *et al.*, 1982). Progoitrins present within rapeseed meal and subsequently converted into goitrins in the intestine, inhibit TMA oxidase. Thus TMA is excreted into the egg yolk (Butler and Fenwick, 1984). This leads to fishy taints in eggs if the concentration of TMA exceeds 0.8 mg/g egg (van Kempen and Jansman (1994). Some of the authors reviewed considered this to be a characteristic of brown egg strains, due to an inability to oxidise trimethylamine. They therefore suggested that the inclusion of rapeseed in feeds for brown egg strains should be restricted. This is consistent with the earlier observations of Overfield (1975).

Similar problems with off-flavours associated with feeding rapeseed meal have been found for poultry meat. For example, Yule and McBride (1976) found adverse effects on meat flavour in broilers fed diets containing rapeseed meal in excess of 50 g/kg.

There has been much genetic selection for major differences between rapeseed varieties. The double low (00) glucosinolate species are considered to be important sources of protein (van Kempen and Jansman, 1994), but their value is affected by the concentrations of phenolics, tannins and phytic acid. Glucosinolates are particularly important since they interfere with thyroid activity in animals and can cause liver haemorrhage in laying hens. Recent cultivars contain 20-30 mmol/g glucosinolates in the oil free dry matter, compared with 150-180 mM/g for high glucosinolate varieties, according to literature reviewed by van Kempen and Jansman (*loc. sit.*). In double low varieties only the following out of the 27 rapeseed glucosinolates are of quantitative importance: progoitrin, gluconapin, glucobrassicanapin, napoleiferin, glucobrassicin and neoglucobrassicin (van Kempen and Jansman, 1994).

Schone *et al.,* (1993) found that soaking rapeseed meals in aqueous myrosinase or copper sulphate solution, and subsequent drying, reduced the content of ANFs by more than 90%. Because glucosinolates and their degradation products are iodine antagonists they tested the treated meals with and without iodine supplementation. Such treatment of the raw material is not likely to be acceptable in organic farming.

Khan *et al.,* (1998) found some evidence that untreated rapeseed meal was associated with an increase in the weight of the thyroid gland of broilers. The effect was prevented by solvent extraction, autoclaving or ferrous sulphate treatment.

High tannin levels in rapeseed meal have been associated with an increased incidence of leg abnormalities (Wise, 1989).

Kastrati *et al.,* (1997) found a progressive decline in rate of lay when feeding increasing inclusion rates of a high glucosinolate rapeseed (125 mmols/g) from 40 up to 120 g/kg.

Khattak *et al.,* (1996) fed a low glucosinolate rapeseed (14 mmols/g) to broilers and found indications that 300 g/kg was an acceptable inclusion rate if the diets were supplemented with vitamin E.

LINSEED (*Linus usitatissimum* L.)

Linseed meal appears to contain an array of ANFs according to literature reviewed by van Kempen and Jansman (1994). It is also low in methionine and lysine. ANFs include a vitamin B_6 antagonist (linamatine), phytic acid, a trypsin inhibitor and cyanogenic glucosides. An inclusion rate of 200 g/

kg depressed broiler growth and eggs from layers fed 150 g/kg linseed had lower preference scores, due to a fishy taint. Ajuyah *et al.,* (1991) found that 200 g/kg depressed broiler performance substantially, though it enriched the carcasses with *n-*3 polyunsaturated fatty acids (PUFA).

Although linseed does not appear to be a valuable protein source for poultry production its potential feeding value is enhanced by its ability to manipulate tissue and egg concentrations of a linolenic acid, docosahexaenoic acid (DHA, 22:6 *n-*6) and eicosapentaenoic acid (EPA, 20:5 *n-*3).

In poultry the fatty acid composition of the tissue lipids is markedly affected by dietary fatty acids, including linoleic and a linolenic acid. These fatty acids can be converted by the bird into longer, more unsaturated fatty acids such as EPA and DHA. EPA and DHA can be supplied in the diet directly as fish oil or as a component of fish meal. Although at the time of writing fishmeal is a permitted ingredient in organic rations (EC 2092/91 and Document 9104/99 ADD 1) there may be consumer preferences for vegetable derived diets. The deposition of a linolenic acid in muscle tissue is advantageous since it can lower the 18:2 to 18:3 (*n-*6:*n-*3) ratio in the human diet, which should increase EPA and DHA synthesis in humans. A ratio of *n-*6:*n-*3 of 5 or less recommended for the human diet (British Nutrition Foundation, 1992; Department of Health, 1994).

Linseed is rich in a linolenic acid compared with rapeseed and soyabean, the typical concentrations being 500 g/kg for linseed versus 120 g/kg for rapeseed and soyabean (Enser, 1999). Enser (*loc. sit.*) when reviewing the work of Ajuyah *et al.,* (1991), Ahn *et al.,* (1995) and Lin *et al.,* (1989) found discrepancies in breast muscle concentration of a linolenic acid and DHA even though the authors had fed diets containing 25-28 g/kg α linolenic acid. Based on the findings for control birds the author suggested that 26 g/kg of dietary α linolenic acid approximately doubled the tissue concentration of EPA and DHA, whereas increases in tissue α linolenic acid concentration varied between twofold and tenfold. The effect of the 18:2 to 18:3 ratio and the Σn-6:Σn-3 ratio of feeding linseed or its oil indicated a marked improvement in human nutritional value. At dietary levels of 26 g/kg α linolenic acid the *n-*6:*n-*3 ratios were below 4.3 in both the white and the dark meat and the Σn-6:Σn-3 ratios were below 2, these being within the recommended range for human nutrition.

The potential to manipulate poultry meat quality, including its eating value, by nutritional means is becoming an increasingly important option for UK poultry meat producers as it offers access to niche markets. Therefore this is discussed further, and in more detail, in later chapters.

Eggs having a high *n*-3 PUFA concentration are already available to the consumer. Reviews on the dietary manipulation of fatty acids so as to increase *n*-3 PUFA concentrations of eggs are available (Hargis and van Elswick, 1993; Leskanich and Noble, 1997; Noble, 1999). As for the organic poultry meat situation there is the potential to produce organic *n*-3 PUFA rich eggs but there may be additional considerations for the organic producer. Although the monetary value of conventional spent hen meat is very low the value of organic spent hen meat may be greater when organic ready meals become more widely available. Enser (1999) warned that if the production of eggs with raised levels of *n*-3 PUFA becomes significant allowance will need to be made for the potentially low oxidative stability of the meat from spent hens.

OILSEED MEAL

Oilseed meals are by-products of the oil food industry. As raw materials they are low in oil content, as the oil has been removed, but they are high in protein content. (Larbier and Leclerq, 1992). The protein is unbalanced because of its vegetable origin and bio-availability and amino acid composition are extremely variable. Carbohydrate is the predominant constituent. This consists essentially of insoluble non-starch polysaccharides, which are components of the plant cell wall and seed hull. These do not present major problems in birds. Larbier and Leclercq (1992) stated that they act as dietary diluents as they are completely indigestible and do not interact with the remainder of the diet. Oilseed meals contain small quantities of ANFs.

NAKED OATS (*Avena* sp.)

The low metabolisable energy value of whole oats (10.5 - 10.9 MJ/kg) has excluded it from broiler and turkey diets (Scott *et al.*, 1982). The protein of oats has high biological value, its amino acid composition is constant over a wide range of protein. A major disadvantage of *A.sativa* is the presence of the hulls, which reduces the feeding value for monogastrics. *A. nuda* L. is a large, naked (huskless) oat species with the caryopses loose within the lemmas and paleas that are alike in texture; hence the grain is similar to that of wheat (Stanton, 1961). Maurice *et al.,* (1985) examined the chemical composition and nutritive value of naked oats in broiler diets. The crude protein content of naked oats used was 172.4 g/ kg. Lysine, methionine + cystine and tryptophan concentrations are given in Table 5. Lysine contents were lower than in soyabean meal.

The authors discussed the apparent inverse relationship between protein content and protein quality. As the protein content of the kernel increases this is associated with an elevation in prolamine content and a consequent decline in protein quality. Avenin (the prolamine in oats) constitutes a small proportion of the protein, this being within the range of 12-20%, whereas in other cereals like wheat and maize prolamine contributes between 30 and 60% of the protein (Mossé, 1966). Maurice *et al.,* (*loc. sit.*) reported an avenin content of naked oats used in feeding trials of 9.4%. Thus oats have an advantage over other cereal grains, and have excellent potential for exploitation as a feed grain because strains with protein contents as high as 35% have been identified (Frey *et al.,* 1975).

In the work of Maurice *et al.,* (1985) naked oats were used to replace maize and soyabean in rations for broilers at a concentration of up to 400 g/ kg. An additional diet was used and this contained 690 g/kg naked oats which replaced all of the maize. Complete substitution of maize with naked oats depressed performance and bone strength but inclusion of naked oats at up to 400 g/kg in corn soyabean based diets did not depress growth. A growth depression at very high inclusion rates was thought to be due to phosphorus deficiency and a relatively high phytic acid content of the   oat-soyabean diet.

Cave and Burrows (1985) used naked oats to substitute for maize and part of soyabean meal, tallow and wheat in diets to provide feeds equal in true metabolisable energy and lysine contents. They used up to 600 g/kg naked oats. The diets were fed to broilers between 28 and 48 days of age. Growth was not depressed by very high inclusion levels of naked oats but feed conversion efficiency adjusted to equal feed intake was poorer when feeding 600 g/kg naked oats. An inclusion rate of 300 g/kg naked oats produced a similar feed conversion efficiency as diets containing none.

Mortality and grading of chicken carcasses were not affected by dietary treatment.

Edney *et al.,* (1989) suggested that the inclusion of naked oats in broiler diets may be limited by the β-glucan content. For example, b-glucanase supplementation in diets containing between 596 g/kg and 666 g/kg oat groats improved growth and feed conversion efficiency between day old and three weeks of age. No response to dietary b-glucanase supplementation was evident for chicks given wheat diets. Amino acid utilisation and starch utilization in chicks fed oat groats were improved by β-glucanase supplementation.

The potential use of naked oats as a home grown protein source in poultry rations is exciting in the context of organic farming. The Organic Food and Farming Report by the Soil Association (1998) identified oats as being the second most widely grown arable crop on organic farms in 1997, the crop being used both for human consumption and livestock feed. It is not known whether naked oats are produced organically. However, work is needed to identify its suitability as a replacement for soyabean meal in wheat based poultry rations as used in the UK.

OTHER ALTERNATIVE VEGETABLE PROTEIN SOURCES

There has been limited work on the use of tomato residues and potato residues as feed ingredients for poultry and it was reviewed by El *et al.,* (1982). Although tomato seed meal, dried tomato pomice and tomato seed cake have reasonable protein contents (410 g/kg, 241 g/kg and 245-370 g/kg, respectively) they are generally lacking in one or more essential amino acid. However, tomato products are a rich source of the carotenoid lycopene (Adams, 1999). There are two principal groups of carotenoids. The carotenes, including lycopene, are hydrocarbons, and the xanthophylls contain carbon, hydrogen and oxygen (Coultate, 1996). Lycopene has 11 conjugate double bonds, and hence its antioxidant potential. Carotenoids in foods have been associated with enhancement of immune response (Adams, 1999).

Xanthophylls have been widely used in the poultry industry to colour egg yolks and the skin of poultry meat and this is achieved through incorporation in the feed. Carotenes including lycopene, are not thought to be effective in colouring animal products.

El *et al.,* (1982) cited the work of Morrison (1936) and Esselen and Fellers (1939) in which they identified dried tomato pomace as being a good source of Vitamin B_1 and a reasonable source of vitamin B_2 and vitamin A.

The work reviewed by El *et al.,* (1982) was largely carried out in the 1920s and 1930s and it would not have taken into account the presence of ANFs.

It is likely that as organic tomato production and processing increases there will be a need to make better use of the by-products. The use of tomato by-products in organic poultry rations may provide a useful outlet for these waste products, but more information would be useful.

Discussion

All of the ingredients reviewed contain impressive lists of ANFs, and it is therefore hardly surprising that their use in the poultry industries of Europe has been viewed with considerable caution. Treatments such as micronisation, pelleting and dehulling seem to have been successful in improving digestibility and in reducing the potency of many of the ANFs, but normally at the expense of changing the content or availability of amino acids. Thus, while it is well known that the legumes are generally low in methionine, micronised peas need to be supplemented with lysine because micronisation has been reported to reduce their lysine content (Igbasan and Guenter, 1996a). Plant breeders have made considerable progress in recent years in the reduction of ANFs, and examples include the double zero low glucosinolate rapeseeds.

There are several serious interpretational difficulties in a review of these ingredients. Firstly standard published tables of nutrient concentrations, such as those quoted in Table 2, tend to be confusing because of the many different versions, names and descriptions of the ingredients, particularly rapeseed.

Secondly there is inevitably a great deal of inconsistency in the results of feeding trials between authors, even those publishing at about the same time. These inconsistencies probably arise because of factors such as differences in varieties used, methods of crop cultivation, environmental differences and the treatments of the seeds after harvest. These factors could have led to differences in the amounts of ANFs, in the amino acid balance and energy content, or even in micronutrient content. For example Igbasan *et al.,* (1997b) showed that pea cultivars varied widely in protein content, from 208 to 264 g/kg, and in starch content from 385 to 437 g/kg. Castell *et al.,* (1996) demonstrated large variations in the composition of peas between cultivars and growing sites.

Thirdly, probably for similar reasons, there are inconsistencies in recommended maximum inclusion levels. Such reasons suggest that the determination of a universally applicable recommendation for inclusion level is impossible. Recommendations must be accompanied by clear definitions of the type of products upon which they are based.

A fourth, more subtle, difficulty encountered in the interpretation of recommended maximum inclusion levels is the variation between authors in their statistical methods. Most have searched for significant differences between treatments, even in work involving multiple levels of inclusion, whereas a more statistically sensitive approach uses response curve analysis (Dillon, 1977; Curnow, 1986). Thus hardly any of the data reviewed above lends themselves to analysis of the marginal cost benefits of increasing ingredient inclusion rate *versus* small performance depressions, all interpreted in the light of the relative prices of eggs and poultry meat *versus* the price of feed as affected by ingredient inclusion rate. The exception is the paper by Castenon and Perez-Lanzac (1990), though even that work used only three inclusion levels and assumed linear responses. Igbasan and Guenter (1997b) used four levels of each of three cultivars fed to white leghorn hens, but did not fit response curves.

Finally, there are particular problems associated with table bird production that have not been encountered before. This is because EU legislation for organic livestock production (EC 2092/91 and Document 9104/99 ADD 1) dictates a minimum growing period of 81 days for table birds or a 70 day growing period if recognised slow growing strains are used. This is vastly different to the duration of the growing period now seen in conventional broiler production, which is typically between 40 and 45 days post hatch.

The intensification of broiler production was stimulated by a need to increase protein supply for human consumption following the World War II and at an affordable cost to the consumer. Breed selection and a better understanding of the hybrids' environment and nutritional needs all contributed to improvements in feed conversion efficiency and a shortening of the growing period. Nowadays there are only a few hybrids that are commonly used by commercial broiler producers and not surprisingly these have widely been used in broiler research. Only limited research has been carried out in recent years on slow growing table birds.

The gross discrepancies between conventional broiler production and organic table bird production, including access to pasture and fluctuating environment conditions, mean that it is not sufficient to draw on research findings established using broilers.

Very little information is available when formulating rations that aim to provide an optimal nutrient intake during growth and achieve a desired market live weight at 70 to 81 days of age. Furthermore, when growing birds to 81 days of age there may also be the need to consider differences between breeds in the rate of development of sexual maturity and the effect of this on their nutrient requirements.

Work is on-going at ADAS Gleadthorpe (MAFF-funded project OF0153) to characterise growth and carcass quality in several slow growing strains, and this includes commercially available hybrids and indigenous pure breeds. In the context of this review slow growing means a growth rate that is less than that achieved by hybrids commonly used in commercial broiler production, under the same growing conditions and nutrient density.

MacLeod *et al.,* (1998) reported that the lysine requirement for growth expressed as a proportion of the diet, was similar between birds of a relaxed-selection line and modern hybrids. Although the absolute lysine intake has greatly increased when comparing modern hybrids and a relaxed-selection line, it has done so because feed intake has kept pace with growth. An economic optimum level of performance was achieved at a dietary lysine content of 11 g/kg.

This may not be true for specific aspects of growth and development, such as feather growth and development where breed differences in feathering may occur. Deschutter and Leeson (1986) suggested that there is evidence pointing to differences in amino acid requirements between strains having genetically distinct rates of feathering. Protein is the major component of feathers (Fischer *et al.,* 1981) and the primary amino acids involved in the synthesis of feather keratin are the sulphur containing amino acids, cystine and methionine (Wheeler and Latshaw, 1981). However, Deschutter and Leeson (*loc. sit.*) suggested that arginine, valine, leucine, isoleucine, tryptophan, phenlyalanine and tyrosine deficiencies have also been implicated in feathering abnormalities. This is not surprising when viewing published amino acid compositions of feathers from young broiler chickens (Larbier and Leclercq, 1994 drawing on the work of Boorman and Burgess, 1986).

Good feathering is important in organic poultry as heat loss on range due to poor feathering will increase feed energy intake for maintenance of body temperature, and crude protein utilisation efficiency is expected to be poorer. Richards (1977) compared normal and partially feathered birds at temperatures ranging from 15°C to 25°C and found the latter to have a 60% higher metabolic rate. Leeson and Morrison (1978) indicated that feed conversion efficiency in laying hens was significantly correlated with feather cover.

Peas seem to be the most promising material. Therefore further statistically sensitive response curve analysis work on multiple levels of the most promising varieties of peas for table birds and brown egg layers, followed by modelling of the responses in a way permitting cost inputs, would be useful. Experiments incorporating at least six concentration levels per pea type, ranging from well below commercial inclusions to well above, would be a way forward. The results would not offer universal recommendations, but would at least provide some guidance for the particular type, or types, of peas used. The pea growing industries will need to provide products of consistent and known quality, with appropriate quality assurance, if the crop is to be widely used by poultry industries.

It is interesting that there has been a concentration of work on peas in France and in Canada, and on lupins in Spain. In view of the thoroughness of the work on peas in France, reviewed in depth and at length by UNIP-ITCF (1995), it is recommended that in scanning the suggested inclusion levels for peas in Table 5 the most weighting be given to the UNIP-ITCF figures.

In the future naked oats may be a valuable feed ingredient and protein source in organic rations but as already suggested work is needed to establish its feeding value for table birds and laying hens under UK conditions.

Linseed may not be a promising alternative protein source but its potential to manipulate fatty acid composition of the muscle tissue enhances its value as a feedstuff.

Whilst solvent extraction of many oilseed products will not produce a permitted feedstuff for inclusion in organic rations the potential to reduce oil contents using an expeller process may be a future option. Conventional linseed cake produced using an expeller process is available. Developments such as this will be dependent on processing efficiency, value of feedstuffs and the size of the market.

The advantages and disadvantages of each of the vegetable protein sources reviewed in terms of their suitability for including in organic poultry rations has been summarised in Table 6.

Table 6. THE ADVANTAGES AND DISADVANTAGES OF VEGETABLE PROTEIN SOURCES REVIEWED IN TERMS OF THEIR SUITABILITY AS AN INGREDIENT IN ORGANIC POULTRY RATIONS

Ingredient	Comments
Peas	Peas of white flowered varieties are a good alternative protein source to soya. They have a relatively high lysine content but the concentrations of sulphur containing amino acids and tryptophan are low. They contain fewer ANFs than soya bean meal and their TIA is less. Micronisation allows peas to be included in rations at higher concentrations but after micronisation it is necessary to supplement with lysine. The metabolisable energy content is compatible with most poultry rations. Oil contents are low.
Lupins	Modest levels of sweet lupins may be included in poultry rations. Typically, they have a similar crude protein content to that of full fat soya. The lysine content is high and the methionine + cystine content is moderate. The quality of sweet lupins as a feed ingredient is variable. They contain lower concentrations of some ANFs than bitter varieties.
Beans	Beans have a moderate crude protein content. The lysine content is high but they are low in sulphur containing amino acids and linoleic acid. The latter is especially important in egg production. Beans have low TIA and lectin concentrations relative to soya bean meal but the concentration of tannins may be high.

Table 6. Contd.

Ingredient	Comments
Sunflower	Sunflower seed meal is a good alternative protein source. Although it is deficient in lysine it is very rich in sulphur containing amino acids. Its inclusion rate in organic diets for laying hens may be limited less by its ANFs than by its moderate metabolisable energy. TIA is low and lectins have not been detected but there are several other ANFs
Rapeseed	Rapeseed meal has a high crude protein content. The lysine and methionine contents are comparable with those of soya bean meal. The metabolisable energy value of rapeseed meal is low but its inclusion rate in poultry rations is limited by the presence of high concentrations of ANFs.
Linseed	Linseed meal has a moderate crude protein content relative to soya but it is low in lysine and methionine. Linseed contains an array of ANFs. Problems with fishy taint and rancidity of poultry meat occur at modest inclusion levels. However, its potential feeding value is enhanced by its ability to influence tissue and egg concentrations of α linolenic acid, DHA and EPA. This is consistent with the Department of Health recommendations on human diet.
Oilseed meals	The crude protein contents are typically high but the amino acid composition and their availabilities are variable. They contain only small quantities of ANFs.
Naked oats	There is the potential to produce naked oats that have a high crude protein content. Lysine content is low relative to soya bean meal but the concentration of sulphur containing amino acid is reasonable. β-glucan content may limit its inclusion rate in poultry rations. More information is required to establish its suitability as a soya replacement in wheat-soya based diets as commonly used in the UK.
Other alternative protein sources	There is an indication that tomato by-products may be suitable feed ingredients in organic poultry rations. This is based on limited analyses of some by-products. The crude protein and lysine content of tomato seed meal is similar to soya bean meal but the concentration of sulphur containing amino acids is low. The potential feeding value may be enhanced by its high antioxidant capacity and benefits this may confer to the birds' level of immunocompetance.

References

Adams, C.A. (1999). *Nutricines*. Nottingham University Press, Nottingham, UK

Ahn, D.U., Wolfe, F.H. and Sim, J.S. (1995). Dietary μ linolenic acid and mixed tocopherols and packaging influences on lipid stability in broiler chicken breast and leg muscle. *Journal of Food Science* **60**: 1013-1018 (Abstract)

Askbrant, S.U.S. (1988). Metabolisable energy content of rapeseed meal, faba bean meal and white flowered peas determined with laying hens and adult cockerels. *British Poultry Science* **29**: 445-455

Ajuyah, A.O., Fenton, T.W., Hardin, R.T. and Sim, J.S. (1993). Measuring lipid oxidation volatiles in meat. *Journal of Food Science* **58**: 270-273 (Abstract)

Ajuyah, A.O., Lee, K.H., Hardin, R.T. and Sim, J.S. (1991). Changes in the yield and in the fatty acid composition of whole carcass and selected meat portions of broiler chickens fed full-fat oil seeds. *Poultry Science* **70**: 2304-2314

Baniel, A., Gueguen, J., Bertrand, D. (1992). Variability of protein composition in pea seeds by high performance liquid chromatography. In: *Premiere Conference européenne sur les protéagineux*. Edited by Angers, A.E.P., pp. 409-410 (Abstract)

Barrier-Guillot, B., Metayer, J.P., Grosjean, F. and Peyronnet, C. (1995). Feeding value of pea presented in mash or pellets in adult cockerels, laying hens, broilers and turkey poults. 10th European Symposium Poultry Nutrition, Antalya (Abstract)

British Nutrition Foundation (1992). Unsaturated fatty acids: nutritional and physiological significance. Report of the British Nutrition Foundation Task Force. Chapman and Hall, London, UK

Brenes, A., Marquardt, R.R., Guenter, W. and Rotter, B.A. (1993). Effect of enzyme supplementation on the nutritive value of raw, autoclaved, and dehulled lupins (*Lupinus albus*) in chicken diets. *Poultry Science* **72**: 2281-2293

Baudet, J. (1983). Compte rendu sur les analyses d'acides aminés et d'azote des pois. INRA-ITCF.FORMA, Versailles, France (Abstract)

Butler, E.J. and Fenwick, G.R., (1984). Trimethylamine and fishy taint in eggs. *World's Poultry Science Journal* **40**: 38-51

Carré, B., Escartin, R., Melcion, J.P., Champ, M., Roux, G. and Leclercq, B. (1987). Effect of pelleting and associations with maize or wheat on the nutritive value of smooth pea seeds (*Pisum sativum*) in adult cockerels. *British Poultry Science* **28**: 219-229

Carré, B. Beaufils, E. and Melcion, J.P. (1991). Evaluation of protein and starch digestibility and energy value of pelleted or unpelleted pea seeds from winter or spring cultivars in adult and young chickens. *Journal Agricultural and Food Chemistry* **39**: 468-472 (Abstract)

Carouée, B. and Duchene, E. (1993). Teneur en protéines du pois. Comment maitriser les variations. *Perspective Agricoles* **183**: 75-81 (Abstract)

Castanon, J.I.R. and Perez-Lanzac, J. (1990) Substitution of fixed amounts of soyabean meal for field beans (*Vicia faba*), sweet lupins (*Lupinus albus*), cull peas (*Pisum sativum*) in diets for high performance laying leghorn hens. *British Poultry Science* **31**: 173-180

Castell, A.G., Guenter, W. and Igbasan, F.A. (1996). Nutritive value of peas for non ruminant diets. *Animal Feed Science and Technology* **60**: 209-227

Cave, N.A. and Burrows, D. (1985). Naked oats in feeding the broiler chicken. *Poultry Science* **64**: 771-773

Charles, D.R., Rennie, J.S., Walker, A.W., Tucker, S.A. and Gordon, S.H. (1996). *Calcium and phosphorus for laying hens.* Unpublished ADAS review.

Conan, L., Barrier-Guillot, B., Widiez, J.L. and Lucbert, J. (1992). Effect of grinding and pelleting on the nutritional value of smooth pea seed in adult cockerel. In: *Premiere conference europeenne sur les proteagineux.* Edited by Agers, A.E.P., pp. 479-480 (Abstract)

Conan, L. and Carre, B. (1989). Effects of autoclaving on metabolisable energy value of smooth pea seed in growing chicks. *Animal Feed Science and Technology* **26**: 337-345 (Abstract)

Coultate, T.P. (1996). *Food. The Chemistry of its Components.* Royal Society of Chemistry, Cambridge, UK

Curnow, R.N. (1986). The statistical approach to nutrient requirement. In: *Nutrient requirements of poultry and nutritional research.* Edited by Fisher, C. and Boorman, K.N., Butterworths, London, UK, pp. 79-89

Department of Health (1994). *Nutritional aspects of cardiovascular disease.* Report on Health and Social Subjects no. 46 HMSO, London

Deschutter, A. and Leeson, S. (1986). Feather growth and development. *World's Poultry Science Journal* **42**: 59-267

Dillon, J.L. (1977). *The analysis of response in crop and livestock production.* Pergamon, Oxford, UK

Dun, P. (1973). Field beans in rations for laying fowls. *ADAS Gleadthorpe Poultry Booklet 1973*, pp. 33-36

Dun, P. (1974). Field beans in rations for laying fowls. *ADAS Gleadthorpe Poultry Booklet 1974*, pp. 52

Dun, P. (1975). Field beans in rations for laying fowls. *ADAS Gleadthorpe Poultry Booklet 1975*, pp. 55

Dun, P. and Hopkins, J.R. (1977). Lupins in a layers diet. *ADAS Gleadthorpe Poultry Booklet 1977*, pp. 47

EEC No 2092/91 (1997). Organic production of agricultural products and indications thereto on agricultural products and foodstuffs to include livestock production (COM(960366 - C4-0481/96 - 96/0205(CNS)) Doc_EN\PR\315\315153

EEC No 2092/91 (1999). Proposal for a Council Regulation (EC) supplementing Regulation (EEC) No. 2092/92 on organic production of agricultural products and indications thereto on agricultural products and foodstuffs to include livestock production, 9401/99 ADD 1

Edney, M.J., Campbell, G.L. and Classen, H.L. (1989). The effect of b-glucanase supplementation on nutrient digestibility and growth in broilers given diets containing barley, oat groats or wheat. *Animal Feed Science Technology* **25**: 193-200

Enser, M. (1999). Nutritional effects on meat flavour and stability. In: *Poultry Meat Science*. Poultry Science Symposium Series Volume 25, Edited by Richardson, R.I. and Mead, G.C., CABI Publishing, Wallingford, UK, pp. 197-216

Ferrez de Oliveira, M.I. and Acamovic, T. (1996). The effect of carbohydrase supplementation on lupin seeds for poultry. *British Poultry Science* **37** (Supplement): S51

Ferrez de Oliveira, M.I. and Acamovic, T. (1999). Apparent digestibility and endogenous amino acids in birds fed enzyme treated and untreated *L.angustifolius* diets. In: *Proceedings of the World's Poultry Science Association Annual Spring Meeting*, Scarborough, pp. 80-81

Fischer, M.L., Leeson, S., Morrison, W.D. and Summers, J.D. (1981). Feather growth and feather composition of broiler chickens. *Canadian Journal of Animal Science* **61**: 769-773

Frey, K.J., McCarty, T. and Rosielle, A. (1975). Straw-protein percentages in *Avena sterilis*. *Livestock Crop Science* **15**: 716-718 (Abstract)

Gilbert, G., Acamovic, T. and Bedford, M.R. (1999). The effect of enzyme supplementation on the growth and feed conversion efficiency of broiler chicks on lupin-based diets. In: *Proceedings of the World's Poultry Science Association Annual Spring Meeting*, Scarborough, pp. 96-97

Gladstones, J.S. (1970). Lupins as crop plants. *Field Crop Abstracts* **23**: 887-1837 (Abstract)

Grosjean, F. (1985). Combining peas for animal feed. In: *The pea crop.* Edited by Hebblethwaite, P.D., Heath, M.C. and Dawkins, T.C.K., Butterworths, London, UK, pp. 453-462

Helsper, J.P.F.G. (1998). Recent developments in improvement of ANF levels in legume seeds by breeding. In: *Proceedings of the Third International Workshop,* Wageningen, The Netherlands, 8-10 July 1998

Hargis, P.S. and van Elswick, M.E. (1993). Manipulating the fatty acid composition of poultry meat and eggs for the health conscious consumer. *World's Poultry Science Journal* **49**: 251-264

Hobson-Frohock, A., Land, D.G., Griffiths, N.M. and Curtis, R.F. (1973). Egg taints: Association with trimethylamine. *Nature* **243**: 304

Huisman, J. and Tolman, G.H. (1992). Antinutritional factors in the plant proteins of diets for non-ruminants. In: *Recent Advances in Animal Nutrition.* Edited by Garnsworthy, P.C., Haresign, W. and Cole, D.J.A., Butterworth Heinemann, Oxford, UK, pp. 3-32

Igbasan, F.A. and Guenter, W. (1996a). The enhancement of the nutritive value of peas for broiler chickens: an evaluation of micronisation and dehulling processes. *Poultry Science* **75**: 1243-1252

Igbasan, F.A. and Guenter, W. (1996b). The feeding value for broiler chickens of pea chips derived from milled peas (*Pisum sativum* L.) during air classification into starch fractions. *Animal Food Science and Technology* **61**: 205-217

Igbasan, F.A. and Guenter, W. (1997a). The influence of micronization, dehulling, and enzyme supplementation on the nutritional value of peas for laying hens. *Poultry Science* **76**: 331-337

Igbasan, F.A. and Guenter, W. (1997b). The influence of feeding yellow-, green- and brown-seeded peas on production performance of laying hens. *Journal Science of Food and Agriculture* **73**: 120-128

Igbasan, F.A., Guenter, W. and Slominski, B.A. (1997). Field peas: chemical composition and energy and amino acid availabilities for poultry. *Canadian Journal of Animal Science* **77**: 293-300

Jansman, A.J.M., Huisman, J. and van der Poel, A.F. (1993). Performance of broiler chicks fed diets containing different varieties of faba bean (*Vicia faba* L.). *Archiv fuer Gefluegelkunde* **57**: 220-227

Kastrati, R., Mestani, N., Domi, Xh., Berisha, B. and Bakalli, R.I. (1997). Effects of dietary high glucosinolate rapeseed meal on the performance and thyroid status of laying hens. *Poultry Science* **76** (Supplement): 81

Khan, M.Z., Mahmood, S., Sarwar, M., Nisa, M. and Gulzar, F. (1998). Effects of various mechanical and chemical treatments of rapeseed meal on the performance of broilers. *Asian-Australasian Journal of Animal Science* **11**: 708-712

Khattak, F.M., Acamovic, T. and Scaife, J.R. (1996). Growth and pigmentation of broilers fed diets containing either whole rapeseed or marine oil with or without vitamin E supplementation. *British Poultry Science* **37** (Supplement): S58-S59

van Kempen, G.J.M. and Jansman, A.J.M. (1994). Use of EC produced oil seeds in animal feeds. In: *Recent Advances in Animal Nutrition*. Edited by Garnsworthy, P.C. and Cole, D.J.A., Nottingham University Press, Nottingham, UK, pp. 31-56

Lacassagne, L. (1988). Alimentation des volailles: substituts au tourteau de soja. INRA *Production Animal* **1**: 47-57 (Abstract)

Lamb, K.J. and Acamovic, T. (1998). The effect of tannin-binding agents, with or without enzyme supplementation, on the dry matter digestibility and ME of Faba beans. In: *Proceedings of the World's Poultry Science Association Annual Spring Meeting*, Scarborough, pp. 75-76

Larbier, M. and Leclercq, B. (1992). *Nutrition and feeding of poultry*. Translated and edited, Wiseman, J. (1994). Nottingham University Press, Nottingham, UK

Lee, D.J.W., Martindale, L. and Paton, I.R. (1982). Rapeseed meal and egg tainting: In *vivo* metabolism and excretion of ^{14}C-trimethylamine by tainter and non-tainter hens. *British Poultry Science* **23**: 175-182

Leeson, S. and Morrison, W.D. (1978). Effect of feather cover on feed efficiency in laying birds. *Poultry Science* **57**: 1094-1096

Leeson, S. and Summers, J.D. (1997). *Commercial poultry nutrition*. University Books, Guelph, Canada

Leskanich, C.O. and Noble, R.C. (1997). Manipulation of the n-3 polyunsaturated fatty acid composition of avian eggs and meat. *World's Poultry Science Journal* **53**: 155-183

Lin, C.F., Gray, J.I., Booren, A.M., Crackel, R.L. and Gill, J.L. (1989). Effects of dietary oils and μ tocopherol supplementation on lipid composition and stability of broiler meat. *Journal of Food Science* **54**: 1457-1460 (Abstract)

Longstaff, M. and McNab, J.M. (1989). Digestion of fibre polysaccharides of pea *(Pisum sativum)* hulls, carrot and cabbage by adult cockerels. *British Journal of Nutrition* **62**: 563-577

MacLeod, G., McNeill, L., Knox, A.I. and Bernard, K. (1998). Comparison of body weight responses to dietary lysine concentration in broilers of two commercial lines and a 'relaxed' selection line. *British Poultry Science* **39** (Supplement): S34

Mannion, A.M. (1995). *Agriculture and environmental change.* Wiley, New York, USA

Maurice, D.V., Jones, J.E., Hall, M.A., Castaldo, D.J. and Whisenhunt, J.E. (1985). Chemical composition and nutritive value of naked oats (*Avena nuda* L.) in broiler diets. *Poultry Science* **64**: 529-535

McDonald, P., Edwards, R.A., Greenhalgh, J.F.D. and Morgan, C.A. (1995). *Animal nutrition.* Longman Scientific and Technical, Harlow, UK

Ministry of Agriculture, Fisheries and Food (1999). *Agriculture in the United Kingdom 1998*, Stationery Office, London

Mossé, J. (1966). Alcohol-soluble proteins of cereal grains. *Fed. Proceedings* **25**: 1663-1669

Mossé, J. (1990). Acides aminés de 16 céréales et protéagineux: variation et clés de calcul de la composition en fonction du taux d'azote des graines. INRA *Production Animal* **3**: 103-109 (Abstract)

National Research Council (1994). *Nutrient requirements of poultry.* National Academy Press, Washington DC, USA

Naveed, A., Acamovic, T. and Bedford, M.R. (1998). Effect of enzyme supplementation of UK grown *Lupinus albus* on growth performance in broiler chickens. *British Poultry Science* **39** (Supplement): S36

Noble, R.C. (1999). Manipulation of the nutritional value of eggs. In: *Recent Developments in Poultry Nutrition 2.* Edited by Wiseman, J. and Garnsworthy, P.C., Nottingham University Press, Nottingham, UK, pp. 251-269

Olver, M.D. and Jonker, A. (1997). Effect of sweet, bitter and soaked micronised bitter lupins on broiler performance. *British Poultry Science* **38**: 203-208

Overfield, N.D. (1975). Egg taints - source of rapeseed meal as a possible factor. *ADAS Gleadthorpe Poultry Booklet 1975*, pp. 54

Richards, S.A. (1977). The influence of loss of plumage on temperature regulation in laying hens. *Journal of Agricultural Science* **89**: 393-398 (Abstract)

RPAN (1993). Nutrition guide. *Rhône Poulenc Animal Nutrition 1993.* (Abstract)

Soil Association (1998). *The Organic Food and Farming report 1998.*

Schone, F., Jahreis, G. and Richter, G. (1993). Evaluation of rapeseed meals in broiler chicks: effect of iodine supply and glucosinolate degradation by myrosinase or copper. *Journal Science of Food and Agriculture* **61**: 245-252

Scott, M.L., Nesheim, M.C. and Young, R.J. (1982). In: *Nutrition of the chicken*. M.L. Scott and Associates, Ithaca, NY, USA (Abstract)

Stanton, T.R. (1961). Classification of *Avena*. In: *Oats and oat improvement*. Edited by Coffman, F.A., American Society of Agronomy, Madison, WI, USA, pp. 75-96

UNIP-ITCF (1995). *Peas: utilisation in animal feeding*. Ed. Carrouée, B. and Gatel, F. (Translated Wiseman, J.) Interprofessional National Union for Protein Rich Crops and Technical Institute for Cereals and Forages. Paris, France

United Nations (1998). *State of the World Population*.

Wheeler, K.B. and Latshaw, T.D. (1981). Sulphur amino acid requirements and interactions in broilers during two growth periods. *Poultry Science* **60**: 228-236

Wiryawan, K.G. and Dingle, J.G. (1995). Screening tests of the quality of grain legumes for poultry production. *British Journal of Nutrition* **74**: 671-679

Wise, D.R. (1989). Nutrition-disease interactions of leg weakness in poultry. In: *Recent Developments in Poultry Nutrition*. Edited by Cole, D.J.A. and Haresign, W., Butterworths, London, UK, pp. 303-319

Yule, W.J. and McBride, R.L. (1976). Lupin and rapeseed meals in poultry diets: effect on broiler performance and sensory evaluation of carcasses. *British Poultry Science* **17**: 231-240

4

FACTORS AFFECTING THE QUALITY OF EXTENSIVELY PRODUCED POULTRY MEAT

Summary of Chapters 4, 5 and 6, market background and bird growth characteristics

Chapters 4, 5 and 6 all deal with aspects of poultry meat quality in intensive production. The introduction which follows links the three chapters.

Introduction to Chapters 4 to 6

It is increasingly difficult for UK broiler producers to compete on price with imported poultry meat products. Measures to improve broiler meat quality, food safety and ethical value have increased UK production costs but, because broiler meat is viewed by many as a commodity food, unit price is a primary criterion to major buyers.

Broiler meat looks similar and tastes similar regardless of where it has been produced. However, in the last few years there has been a food revolution within the UK, and increasing numbers of consumers are now seeking food products with enhanced flavours and nutrient contents.

Initial demands for full flavoured poultry meat were met by retailers importing Label Rouge birds from France, but UK production is now making a valuable contribution. The niche market is important to economic sustainability. Ideally, there will be visual characteristics which will enable consumers to identify premium table birds.

Broilers are sold whole, and they may be roasted, or portioned and casseroled, grilled, fried, steamed, or barbecued. They are used in many further processed products, where the flavour is achieved through seasoning, spices, or herbs.

Production factors affecting poultry meat quality, including organoleptic properties and nutrient contents, have been reviewed. Growth rate is central to many eating quality characteristics. This means that factors affecting growth rate and live weight at slaughter (e.g. genotype, duration of the growing period, diet specification, and ambient temperature), influence meat flavour and texture, carcass conformation and nutrient content. Broiler hybrids have been selected for fast early growth so that they reach typical market live weights of about 2.0 to 2.2 kg at a young age, and the meat is tender enough for fast cooking at high temperatures. Their use in free range production systems, where the minimum age at slaughter is day 56, is acceptable, especially if only females are used, since they gain weight more slowly than the males. Furthermore, females have an earlier growth inflexion than the males and meat flavour is enhanced post growth inflexion. This is due to the deposition of flavour precursors in the muscle, although these have not yet been identified. Meat texture of broiler hybrids grown to day 56 in a free range system is likely to be different from that produced when the birds are grown conventionally. This is because the cross sectional areas of muscle fibres are likely to be smaller in birds fed low specification diets, and therefore, when chewed, the meat fibres may feel more tightly packed.

Broiler hybrids do not have a growth profile suited to 81 day production. Even on low specification diets the birds would need to be feed restricted so as not to exceed market live weights much before day 81. Imported slow growing hybrids are the most suited for 81 day production. They have been used in Label Rouge production since the 1960s. Traditional UK breeds have the potential to be used in speciality 81 day plus production. The growing period is likely to exceed 100 days. This would increase production costs which would have to be passed on to the consumer. The flavour and texture of these birds would have to be outstanding in order to sustain repeat purchase.

Slow growing hybrids slaughtered at day 81 are expected to have firmer meat texture and more flavour than broilers. Again, this is because the cross sectional areas of muscle fibres are likely to be smaller than in those of broilers. As growth inflexion will occur well before the minimum slaughter age for traditional free range birds, their flavour will be enhanced. In the males, flavour intensity may be increasing because the birds are starting to develop sexually and this may lead to flavour variation between birds at the point of sale. One means of counteracting variation in flavour between birds would be to have a documented research-based system for predicting the flavour intensity of males and females according to production factors (e.g. age at slaughter, diet specification, environmental temperatures during growing, increasing or decreasing daylengths). A flavour score may then be applied so that consumers are able to purchase a table bird according to taste preferences.

Most of the work reviewed in these three chapters examined breed differences in meat qualities and eating qualities at live weight, or age reference points, but a more useful research approach may be to compare qualities at similar metabolic ages. If differences between breeds occur at the same metabolic age then this would be due to genotype. An understanding of the processes behind age-related flavour development may enable strategies to be developed that aim to optimise the taste and aroma of cooked poultry meat.

The effects of cooking method and final cooking temperature on the development of aroma in poultry meat has not received a great deal of attention. This may be an increasingly important component of eating quality work, as consumers buying premium products will want "the whole experience", of which cooking and kitchen aromas will be components. Cooking temperature may also influence the gelatinisation of collagen and meat toughness.

Ranging may affect muscle characteristics, and in breast muscles the fibre length may be increased. This may influence meat texture, most notably the perception of fibres, but the relative contribution of exercise to poultry meat quality is not known. Furthermore, some birds may range more than others, and some breeds may be better rangers. There may also be effects of wind, rain and sunshine on ranging. As ranging is likely to be very variable both within and between flocks, meat eating quality may vary.

Pasture intake and the contribution of pasture to the birds' total nutrient intake is likely to be variable, but low. This is because pasture compositions have been developed on the basis of ease of pasture management, and for their hard wearing properties. However, in an extensive production system, the pasture should be considered as being a valuable source of nutrients, and sward compositions should pertain to this ideal. Furthermore, herbage intake may intensify poultry meat flavour, although this will probably be at least partly dependent on intake being sufficient to modify gut microflora.

There is some indication that meat from slower growing birds may be redder than broiler meat, and if this is related in part to a higher meat pH, then the shelf life of traditional free range birds may be shorter than that of broilers. This is because a higher meat pH favours the growth of meat spoilage bacteria. As premium poultry meat is a much more valuable product that commodity broiler meat, and the purchase of premium poultry meat may be more erratic, it is important that the shelf-life of table birds is maximised. Research should be aimed at establishing whether there are differences in meat pH between table birds and broilers, and whether shelf-life is affected. Methods for extending shelf-life should be proposed.

The carcass conformation of table birds will differ from that of broilers in that the ratio of white to dark meat will be reduced. This may be an acceptable feature of whole table birds but, if there is to be an expansion into premium poultry meat portions, then the yield of breast meat will be important. The dietary supply of essential amino acids to broiler hybrids grown slowly to day 56 has probably not been optimised, and so there may be some scope to increase breast meat yields through dietary means. By comparison, breast meat yields at about day 81 in Label Rouge birds seem to be little affected by diet specification. Perhaps breeder companies supplying slow growing chicks to UK producers will see this as an important and increasing market, and if so they may develop strains that meet their new customers' requirements.

MARKET BACKGROUND

It is important to the development and survival of the niche poultry meat market that the product has recognisable and definable traits that set it apart from commodity chicken. Ideally, the consumer will be aware that the birds are grown locally, and that consideration is given to the birds' welfare and environmental issues relating to production, and that the product is of good quality and safe. However, it will be the idea that the product is something different, something richer in flavour, that will drive the purchase.

There is potential to produce a wide range of premium table birds, in terms of appearance, meat flavour and meat texture. As the range of table birds increases, labelling will be important as this is the only means of letting the consumer know where and how the bird has been grown. Differences between production systems (conventional broiler production, free range table bird production and traditional free range table bird production systems) should be clear to the consumer. The main points of difference between production systems are given in Table 1, but this type of detailed information is likely to be meaningless to the consumer. Pictures of birds ranging are more likely to have an impact. Producer brands may be useful, particularly if, through advertising, the product may be tied to a region. Perhaps of most use to the consumer would be a subjective scoring system for describing the birds' flavour intensity and meat texture. A subjective scoring scheme for describing the strength of cheeses has been widely adopted within the UK and it tempts consumers to try speciality products at low risk.

Table 1. MAIN POINTS OF DIFFERENCE BETWEEN CONVENTIONAL POULTRY MEAT PRODUCTIONS SYSTEMS

| | Production system | | |
| | *(a) Intensive* | *(b) Extensive* | |
	Broiler	*Free range table bird*	*Traditional free range table bird*
Minimum age at slaughter (days)	None, but generally between 39 and 45	56	81
Breed specification	None	None	Slow growing
Maximum house stocking density	34.0 kg live weight/m^2	13 birds/m^2 or 27.5 kg live weight/m^2	12 birds/m^2 or 25.0 kg live weight/m^2
Flock size	Unlimited	Unlimited	Maximum 4,800 birds
Access to range	Not required	Continuous daytime access required for at least half their life-time	Continuous daytime access required at least from 42 days of age
Pasture allowance	None	1 m^2 per bird	2 m^2 per bird
Feed specification	None	Finisher contains at least 70% cereals	Finisher contains at least 70% cereals

If a subjective scoring scheme for describing the flavour intensity and texture of poultry meat is to be applied, then knowledge of production factors and the effects on these traits would need to be established. This review highlights some of the potential effects of production on meat flavour and texture.

PRODUCT CONSISTENCY

Although it is thought that there will be a demand for an ever-wider range of table birds, variability within a product will not be desirable. Consumers may be disappointed, if on repeat buying, the birds' flavour is less intense than before, or they may be surprised if the flavour is more intense. However, the very nature of free range production; that birds are able to choose between environments and that environmental conditions will vary daily, and between seasons, mean that product variability is likely to be greater than in broiler production. This may be suitably taken into account if birds are sold according to a given flavour and texture description.

ASPECTS OF QUALITY

There are many attributes that contribute to the overall quality of a table bird. For example, the bird is a desired weight, conformation is good, the carcass is free from blemishes, bruises, breast blisters, hock burn and scratches, skin colour is pleasing and meets the consumers' preference for a white, cream or "corn-fed" bird, the skin finish is good (this relates to the fat content), the skin is not torn, the feathers have been cleanly plucked, the skin "crisps" during roasting or frying, and meat colour, texture, succulence and flavour are good. In addition there is an increasing need to consider the nutrient content of meat, particularly the fat content and its fatty acid composition. Whilst many meat quality characteristics will be determined by production system requirements (e.g. minimum age at slaughter, or diet specification) there is the potential to optimise eating quality through breed choice, management practices and diet.

The aims of these chapters are to examine the effects of length of growing period, breed, management practices and feeding strategies on poultry meat quality, including organoleptic properties and nutrient content. However, it is also important to consider the consumers, who they are and why they may be buying premium poultry meat, as they are the driving force for niche food production.

Who purchases extensively produced poultry meat and why?

In identifying and developing new products for sale it is helpful to view the market from the customer's point of view. Who purchases food and why? What are the forces driving food purchasers and influencing their opinions? This is particularly important when developing premium food products, as the purchaser's expectations will be much higher than for staple foods. Understanding the consumer's preferences for poultry meat, and the appropriateness of the product in terms of lifestyle and cooking method, will be important.

The development of niche markets, including the market for extensively produced poultry meat, is driven by consumer demand. It is estimated that at present extensively produced table birds account for less than 2% of total table bird production within the UK, but growth is expected. Until then, it is possible only to predict who is buying premium poultry meat.

Most market research organisations classify households according to the gross weekly income of the head of the household, and the class intervals given in Table 1 of Chapter 2 (page 16).

Differences in food spending habits exist between the different income groups. For example, eggs in shell are a cheap abundant commodity staple, and OAPs and E1 s spent more per person per week on eggs than classes A and B, and by a large margin (MAFF, 1999). Other bases of classification of consumers include classification by life stage. Charles (2000) suggested that there are seven classes according to life stage, and they are given in Chapter 2, page 16. There are also classifications of consumers based on occupation, postcode and aspirations and attitudes.

Charles (2000) also suggested a classification for motivations driving purchasers (see Chapter 2, Table 2, page 18), but recognised that this may be an over simplification of real life purchases, because few people behave in strictly definable ways. He listed six types of purchases: 1) price motivated purchases, and these were staples that were sold retail to consumers without much alteration from the form in which they left the field or animal; 2) convenience and time saving motivated products, such as ready meals; 3) healthy eating motivated purchases, including fruit, vegetables, high fibre foods, low fat foods, and organic foods; 4) interest, variety and fun motivated purchases, including theme and regional foods; 5) gifts, leisure and pleasure purchases, and; 6) social and emotional food purchases, that is food for family groups, or at business gatherings (see Chapter 2 for more details).

It is suggested that the main relevant purchasers of extensively produced poultry meat will be income groups A and B, particularly when such types are in "leisure mode", as defined by Hughes and Ray (1999), and they are thought likely to be in life stages 3, 5 or 6. Healthy eating, interest and variety, and social dining are thought to motivate the purchase of extensively produced poultry meat.

Consumer satisfaction

Any premium food product failing to meet the consumer's expectations is unlikely to attract repeat purchases. A major problem for producers is that very little is known about UK consumers' organoleptic preferences for poultry meat. The reason for this is that since the 1960s UK consumers have mostly purchased broiler meat, because traditionally grown chickens were not widely available. Long-term exposure to broiler meat texture may cause some consumer resistance to firmer poultry meat, such as generally found in slow growing birds, but the enhanced flavour of older birds may be preferred. One indication of UK consumers' preferences was given in a recent report published by the Food Standards Agency (May, 2000). When addressing public attitudes to food safety through a number of focus groups (which included people from socio-economic groups B, C, D and E, lifestages covering ages 18 to 65, but weighted to respondents with children at home, females and those in urban areas), many respondents thought that chickens nowadays were "beautifully plump", but that they "lacked taste". Almost everyone believed that free range chickens were healthier to eat and tasted better.

"Label Fermier" poultry meat is widely consumed in France (about 31% of chickens bought by French consumers are Label Rouge (Laszcyzk-Legendre, 1999) and many studies have reported French consumers' preferences for the firmer meat and enhanced flavour, compared with broiler meat (e.g. Culioli *et al.,* 1994). Sonayia *et al.,* (1990) described Nigerian consumer preferences for firmer poultry meat as produced in traditional free range systems, whereas broiler meat was too tender for Nigerian cookery methods. By comparison, Japanese consumers preferred meat from young broilers because it was softer and had more flavour when steamed in the presence of salt (Yamashita *et al.,* 1976).

A vital first step for the UK poultry meat industry will be for it to determine UK consumers' organoleptic preferences, as the eating quality of premium poultry meat can be optimised only with reference to this information. At the time of writing this chapter, neither the British Chicken Information Service, nor the British Poultry Meat Federation had any information available on UK consumer preferences for eating quality traits in extensively produced table birds.

There may also be a need to examine cooking methods, as older birds may have different optimum cooking temperatures and cooking times from those used for cooking broiler chickens, or small traditional free range birds. Perhaps one of the first intimations of this was in the popular TV cookery series by Smith (1999) where she proposed optimal methods for cooking birds of different weights. The method proposed for cooking small (1.35 kg) traditional free range chickens, involved fast roasting at a high oven temperature. Larger birds (2.25 to 2.70 kg) were thought to taste best when cooked slowly at a moderate oven temperature. In this case, the breast of the bird was smeared with butter and protected during cooking by positioning streaky bacon on top of the breast. Both acted as bastes during cooking. A pork sausagemeat and liver stuffing was placed inside the body cavity prior to cooking and this provided a means of maintaining moisture and adding fat during cooking.

Growth rate and body composition

Growth rate, body composition and feed intake are inextricably linked (Emmans, 1987). This means that body composition is expected to differ between birds fed *ad libitum* and those which are feed restricted, and between birds fed optimal rations and birds fed less than optimal rations. Thus, a broiler hybrid fed optimal rations and reaching a market live weight of 2.0 kg at day 40 will have a different body composition to that of a growth restricted bird, where the same live weight is achieved, but not until day 56 or day 81.

McMeekan (1940) reported effects of feeding on carcass growth in pigs. He proposed the hypothesis that when nutrient intake is inadequate for optimal growth, a higher priority for nutrients is given to body parts like the head, neck and ears than to commercially important parts of the pig, like the loin and hind quarters. His hypothesis was that skin, tendons and glands were penalised less than skeleton, which was penalised less than muscle, which was penalised less than fat. Hammond (1960), developing the same concept, showed that the plane of nutrition determined the shape of growth curves for a strain, and this had implications for carcass composition in mammals both before and after birth. He stated that the time of the rate of maximum growth of each tissue, or part, occurs in a definite order, and the order determines the priority for nutrients. In pigs, breeds differed in the rate and extent to which proportional growth changes, therefore, optimal ages at slaughter differed (Hammond, 1932).

Growth priority concepts are perhaps less relevant to modern broiler production as diets are usually formulated to be optimal, and feed is usually available *ad libitum*, but they are relevant to extensive table bird production since nutrient intakes are not likely to be optimal. Whole body growth will be considered first, and this will be with respect to meeting the consumers' requirements for size of table bird, followed by a discussion on allometric growth and carcass composition, and then muscle growth will be considered.

WHOLE BODY GROWTH

Differences in breed growth rates and effects of diet specification

There is a requirement to attain a live weight range at market age that meets the consumers' needs and their price range. Typical market live weights are between 2.0 and 2.5 kg. Allowing a killing out percentage of approximately 65% for birds fed Label Rouge type rations (MAFF-funded project OF0153), this will produce dressed birds of between 1.3 and 1.6 kg. Although independent high street butchers usually sell table birds on a price per unit weight basis, supermarkets have tended to adopt a common price for whole chickens in a given weight category. Larger scale producers selling chickens to supermarkets will not only need to produce birds having a suitable market live weight, but they will also need to minimise variability in live weight, so that there is minimal variation around the mean dressed weight.

Slowing the growth of broiler hybrids so that live weights of between 2.0 and 2.5 kg are achieved at a minimum slaughter age of 56 days will not be easy. It will be especially difficult to slow growth sufficiently, to achieve these target weights at a minimum slaughter age of 81 days. This is because commercial broiler stocks have very fast growth rates early in life, and a live weight of 2.0 kg may be reached by day 40. High dietary cereal contents, as required in rations for extensively produced chickens (EC Poultry Meat Marketing Standards, 1997), reduce the dietary crude protein contents compared with broiler rations, and this assists in slowing broiler growth rates, but not sufficiently for 81 day production. Lewis *et al.,* (1997) reported faster early growth rates for Ross 1 broilers fed Label Rouge type rations than for ISA 657 birds (a hybrid widely used in France's 81-day traditional free range production system, and sold as "Label Rouge" table birds), and it was only at day 83 that live weight gains were similar between strains. At day 83, the mean live weight of ISA 657 birds was 2.79 kg, compared with a mean live weight of 4.57 kg in Ross 1 broilers.

An experiment at ADAS Gleadthorpe (MAFF-funded project OF0153) examined growth rates of UK broiler hybrids fed either presumed non-limiting rations (with respect to the Ross 308, commercial broiler), or Label Rouge type rations. As-hatched live weights at days 56 and 81 are shown in Figure 1.

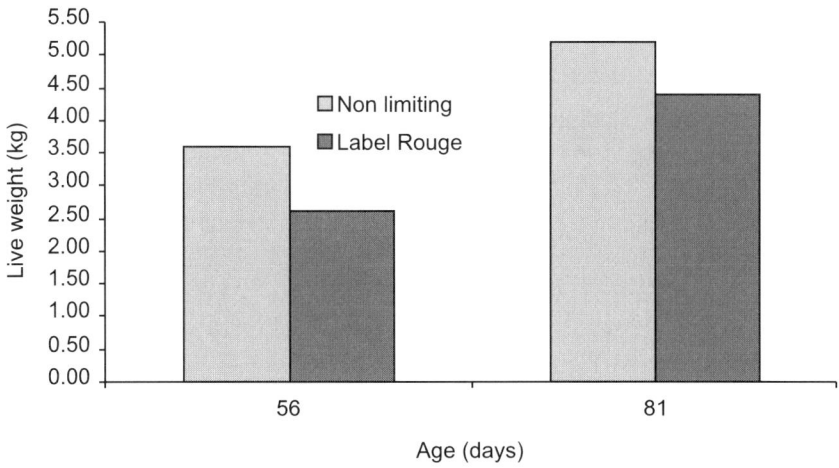

Figure 1. Mean live weights of as-hatched Ross 308 broilers at days 56 and 81 when fed either presumed non-limiting rations, or Label Rouge type rations and grown in a controlled environment with no access to pasture.

The data presented in Figure 1 illustrate that a commercial broiler hybrid fed *ad libitum* has a much heavier live weight at day 81 than that typically required by the consumer, despite the lower nutrient specification of Label Rouge type rations. This suggests that, if commercial broiler hybrids are grown in extensive production systems where the minimum age at slaughter is 81 days, then feed restriction will be necessary in order to prevent them from becoming too heavy.

Feed restriction may not be acceptable to consumers, as birds grown in extensive production systems are expected to have better standards of welfare than intensively produced broilers. Also, if the suggestion by Emmans and Fisher (1986) that animals seek to eat because they seek to grow, and presumably the seeking to grow is driven by the genome seeking to meet its programmed mature body size, then feed restriction may jeopardise bird welfare. Emmans and Fisher (1986) proposed the following equation:

$$DFI = R_1/C_1$$

where, DFI = desired feed intake, kg per day

R_1 = requirements, units per day, for the first limiting resource
C_1 = the feed content, units per kg, of the first limiting resource.

Although it could be argued that birds in extensive systems would have access to pasture, and that intake of pasture is unlimited, fresh grass is very low in protein, lysine and methionine, compared with broiler rations, or Label Rouge type rations (Table 4), and moisture contents and crude fibre contents are much higher in grass. A high crude fibre content would limit grass intake because bulkiness would affect gut fill and digestibility. Low lysine and methionine contents in grass, these being the first and second limiting amino acids for growth (Larbier and Leclercq, 1992), would not assist in reducing the birds' desire to feed as described by the Emmans and Fisher (1986) concept.

Furthermore, when quantitative feed restriction techniques are applied to birds in communal groups, it is likely that not all birds are able to feed at the same time, and some birds will receive less than their share of feed.

Table 4. CRUDE PROTEIN, LYSINE AND METHIONINE CONTENTS OF
TYPICAL BROILER RATIONS, TYPICAL LABEL ROUGE TYPE RATIONS AND
GRASS, GIVEN ON A DRY MATTER BASIS (g/kg)

Feed type	Ration	Dry matter (g/kg)	Crude protein (g/kg)	Lysine (g/kg)	Methionine (g/kg)
Broiler[1]	Starter	880.0	210.0	1.32	0.66
	Grower	880.0	210.0	1.29	0.62
	Finisher	880.0	185.0	1.22	0.58
Label Rouge[1]	Starter	873.0	182.0	1.03	0.37
	Grower	871.0	161.0	0.86	0.31
	Finisher	870.0	159.0	0.76	0.29
Grass[2]	Young	200.0	156.0	-	-
	Mature	282.0	100.0	-	-
	Dried	896.0	-	7.92	3.35

Sources:
[1]MAFF-funded project OF0153
[2]McDonald *et al.,* (1995)

Increased competition at feeding may cause aggression and bird damage,
and even moderately short periods of feed withdrawal have been shown to
cause physiological stress. Karunajeewa (1987), citing the work of Freeman
et al., (1980), reported increased plasma corticosterone concentrations
two hours after the onset of feed withdrawal in three-week old Light Sussex
birds. The birds also developed hypoglycaemia and hyperlipacidaemia. In
later work by Freeman *et al.,* (1981), birds had elevated plasma
corticosterone concentrations for five weeks following feed restriction early
in life, and plasma fatty acid concentrations remained high for eight weeks.
This would mean that restricting feed to broiler hybrids in early life would
cause physiological stress for at least 43 to 69% of the duration of an 81-
day growing period, and feed restriction is thought to be necessary for the
majority of the growing period if target live weights of between 2.0 and 2.5
kg are to be reached at day 81. Lewis *et al.,* (1997) reported that Ross 1
broilers fed so as to achieve a similar live weight at day 83 as ISA 657
hybrids, had feed intakes of only 55% of that of *ad libitum* fed Ross 1
broilers.

It seems that the primary selection criterion for identifying breeds suited for use in free range and traditional free range production systems should be growth rate, and the attainment of a required market live weight no sooner than the minimum slaughter age when fed on an *ad libitum* basis. There are wide differences in growth rates both between commercially available hybrids (mostly stock imported into the UK from France) and between traditional UK breeds. In work at ADAS Gleadthorpe (MAFF-funded project OF0153) growth rates of several commercial hybrids and traditional breeds fed either presumed non-limiting rations (with respect to the Ross 308, commercial broiler), or Label Rouge type rations were examined. As-hatched live weights achieved at days 56 and 81 are shown in Figures 2 to 5.

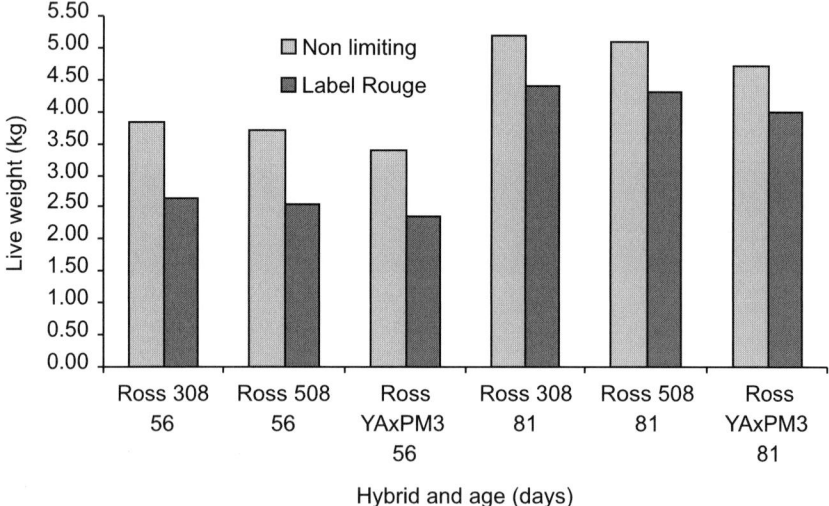

Figure 2. As-hatched live weights at days 56 and 81 for UK commercial broiler hybrids fed either presumed non-limiting rations (with respect to Ross 308), or Label Rouge type rations, when housed in a controlled environment.

Differences in growth rates between commercial hybrids are real and deliberate as the breeder companies have developed hybrids suited for particular markets. For example, the phenomenal growth rates of modern broiler hybrids demonstrate the success of breeder companies in selecting for early growth. This trait, and commensurate improvements in feed conversion efficiency, have provided an abundant and relatively cheap supply of poultry meat for UK consumers (Table 5).

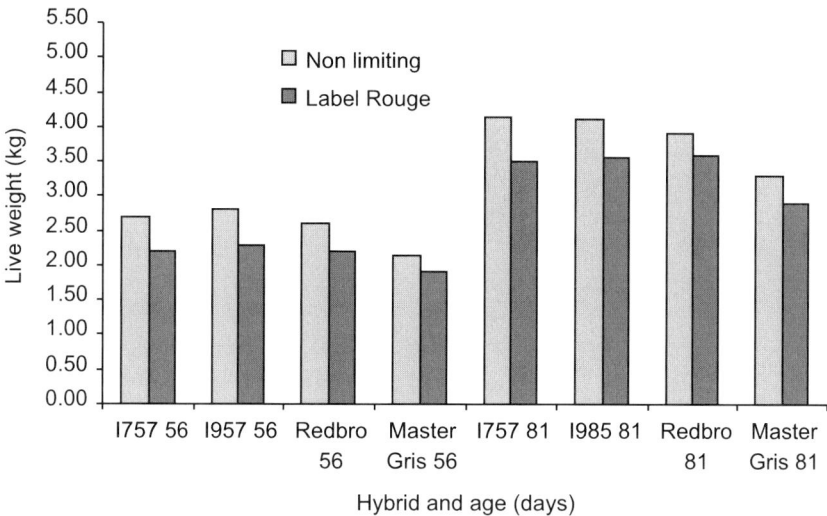

Figure 3. As-hatched live weights at days 56 and 81 for imported fast-moderate growing hybrids (ISA) fed either presumed non-limiting rations (with respect to Ross 308), or Label Rouge type rations, when housed in a controlled environment

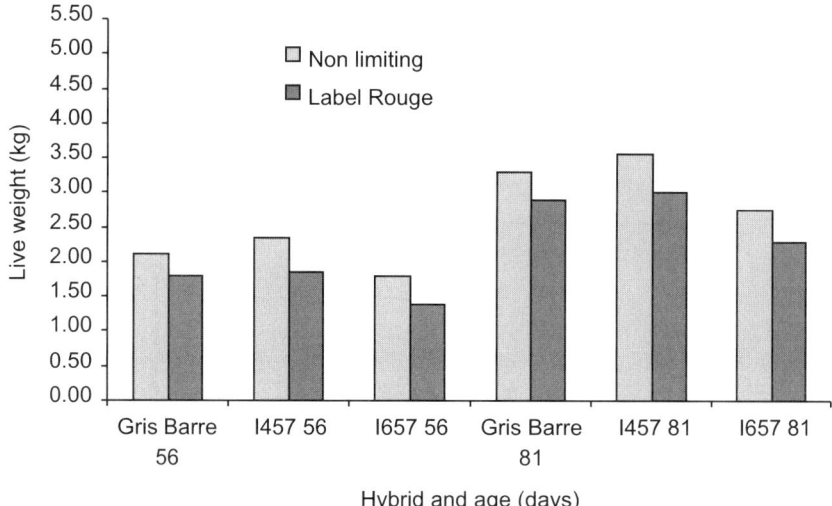

Figure 4. As-hatched live weights at days 56 and 81 for imported moderate-slow growing hybrids (ISA) fed either presumed non-limiting rations (with respect to Ross 308), or Label Rouge type rations, when housed in a controlled environment

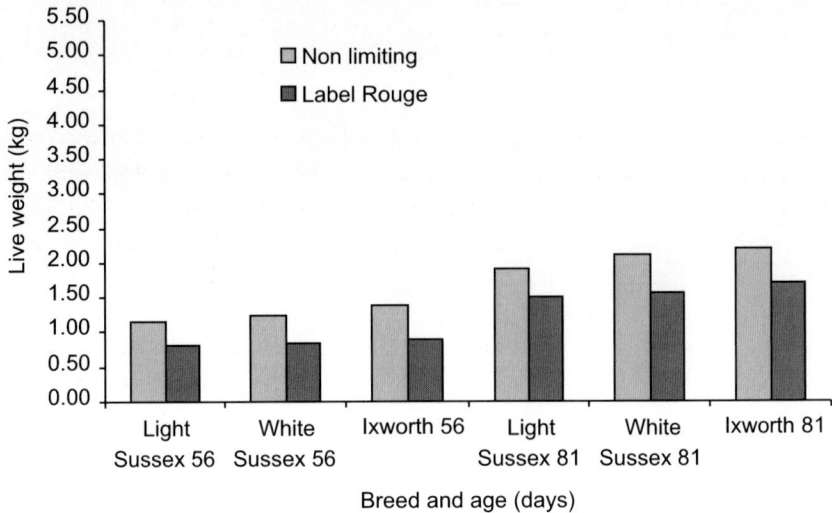

Figure 5. As-hatched live weights at days 56 and 81 for traditional breeds fed either presumed non-limiting rations (with respect to Ross 308), or Label Rouge type rations, when housed in a controlled environment

Table 5. COMMERCIAL BROILER DATA (PERFORMANCE IN JULY OF EACH YEAR)

Year	Killing age (days)	Weight (lbs)	FCR
1971	58	4.00	2.20
1981	52	4.50	2.20
1991	47	4.98	2.10
1995	45	5.10	2.00

Source: Whittle (1998)

Improvements in the growth rate of broilers through time, and the relative contributions of genetic selection and nutrition are illustrated in the work of Havenstein *et al.,* (1994). The authors compared growth rates of the 1991 Arbor Acre broiler with that of the 1957 broiler (Athens-Canadian Randombred Control) when fed either a 1957 diet, or a 1991 diet. The 1957 starter and grower rations had lower crude protein, lysine and methionine contents than the 1991 starter and grower rations, and the energy values were also lower than the 1991 starter and grower rations (e.g. 1957 starter ration had 217 g crude protein/kg, 10.9 g lysine/kg, 4.3 g methionine/kg, and 12.0 MJ ME/kg, compared with 225 g crude protein/kg, 12.1 g lysine/kg, 5.1 g methionine/kg and 12.9 MJ ME/kg in the 1991 starter ration). Live weights of the 1991 broilers at day 56 when fed the 1991 diet were approximately 3.1 times greater than the 1957 broilers, and similarly approximately 2.7 times greater at day 81 (Table 6). Improvements in diet composition since 1957 increased live weights at day 56 by 26.7% in the 1957 broiler, whereas live weights at day 56 were 13.9% poorer when 1991 male broilers were fed the 1957 diet. The 1957 starter and grower rations were not dissimilar to the Label Rouge type rations used at ADAS Gleadthorpe, but in the latter work modern broiler live weights at day 56 were reduced by approximately 31% when comparing birds fed Label Rouge type rations with birds fed presumed non limiting rations.

Table 6. LIVE WEIGHT BY BREED, DIET, SEX AND AGE

Broiler strain	Diet	Sex of bird	Live weight (kg)	
			day 56	day 81
1957	1957	Male	0.858	1.564
1957	1957	Female	0.722	1.236
1957	1991	Male	1.087	1.882
1957	1991	Female	0.893	1.480
1991	1957	Male	2.901	4.579
1991	1957	Female	2.507	3.779
1991	1991	Male	3.368	4.770
1991	1991	Female	2.848	4.226

Source: Havenstein *et al.,* (1994)

Genetic improvements in broiler growth rates over the last forty to fifty years mean that the broiler hybrids of today have too fast a growth rate for achieving typical market live weights of between 2.0 and 2.5 kg at day 81. Even when low specification diets are fed to broiler hybrids, the live weights achieved at day 81 are greater than those commonly wanted by the customer. This means that slow growing birds will need to be used in traditional free range production and, this will have cost implications for UK producers as it will be dearer to use traditional breeds or imported stock, than to use broiler hybrids. There will also be implications for carcass conformation and meat eating qualities and, these will be discussed later in the review.

It is thought that, for the traditional breeds, there has been little co-ordinated effort to improve the birds' growth performance, conformation, or eating qualities. This is because the traditional breeds tend to be kept in very small flocks, and the industry is too small to have a set of clear objectives or organised breeding programmes. It is not surprising, therefore, that it is the French commercial breeder companies that have stock suited to an 81-day growing period. Label Rouge production requires a slow growing bird which reaches a typical market live weight of between 2.0 and 2.5 kg at day 81, and this production system has been operating in France since 1965 (Laszcyzk-Legendre, 1999). In 1998, there were 96.7 million birds produced in the Label Rouge system (Laszcyzk-Legendre, 1999).

Traditional UK breeds offer producers "new" sources of genetic material, and their use in extensive systems of poultry meat production should not be over-looked. Traditional breeds do achieve the required market live weight of between 2.0 and 2.5 kg and heavier, but their growing period would be longer than the 81 day minimum growing period required for traditional free range production. Roberts (1997) cited mature live weights of 4.1 kg for Ixworth males and Sussex males, and 3.2 kg for Ixworth females and Sussex females. The slower growth of traditional breeds, compared with commercial hybrids, means that the production costs of traditional breeds are much likely to be higher, and this would need to be passed on to the consumer. Killing out percentages are also expected to be lower than for commercial broilers (e.g. approximately 60-65%, compared with over 70%, respectively) and this will add to the cost. Organised breeding programmes for the traditional breeds could, perhaps, improve growth rates and feed conversion efficiencies, but this would take time and investment. However, there is the potential to market rare breed poultry meat to consumers who are seeking a product perceived as more authentic and traditional.

Rather than simply examining breed differences in live weight at a given age, a more useful approach is to fit growth functions since they summarise information obtained from a sequence of points (weights or ages) into relatively few parameters. This facilitates a comparison of growth efficiency of the breeds, or hybrids. Growth functions also allow the prediction of weight at a specific age.

The Gompertz function (see also definitions section above) is widely used to describe the growth of chickens (Wilson, 1977; Tzeng and Becker, 1981; Ricklefs, 1985; Knizetova *et al.,* 1991; Lawrence and Fowler, 1997). It is a sigmoid (i.e. S-shaped) equation that was developed by Gompertz (1825) for actuarial purposes, but the function has been applied to growth data since the 1930s (Winsor, 1932). The assumption of the function is that animals, fed *ad libitum* are capable of maximum growth. The three parameters used to describe the growth curve are mature live weight, the rate of maximum live weight gain, and live weight at age zero. In poultry, it is conventional to use live weight at hatching as live weight at age zero, despite a 21 day growth phase during incubation. It is noted that the Gompertz function is a mathematical expression of the growth of the whole organism that is based on the rate of exponential decay with age, rather than a description of precise biological processes.

The S-shaped curve is used to describe the tendency for growth rate to accelerate at first, then to enter an approximately linear phase at a steady pace, and finally to decelerate as animals approach their final mature weight.

Wilson (1977) compared several possible functions fitted to weight data for the Ross 1 male broiler, and the Gompertz function fitted well. Tzeng and Becker (1981), using the Gompertz function, predicted a mature live weight of 4.73 kg for commercial pure bred male broiler selection lines. Peak live weight gain was 64.4 g/day at day 44, then weight gain reduced to 46.8 g/day at 69 days and approached zero as the birds got older. Knizetova *et al.,* (1991) evaluated the Gompertz function for nine broiler lines and observed that the age of inflexion varied between 48 and 56 days for males and between 48 and 53 days for females. Larbier and Leclercq (1992) published growth inflexion points of 48.2 days and 43.2 days for male and female broilers, respectively, and a growth inflexion point of 52.8 days in male Label Rouge birds.

The Gompertz function, as detailed below, has been fitted to as hatched live weight data collected in a study at ADAS Gleadthorpe which examined differences in breed growth rates when fed presumed non limiting rations (with respect to the commercial broiler, Ross 308), or Label Rouge type rations. Derived Gompertz parameters are given in Table 7.

Gompertz function

W = A + C * exp(-exp(-B*(t- M)))

where:

W is the live body weight at time, t (kg)

t = age of the chicken (days)

The parameters A+ C, B and M were interpreted as follows:

A + C = the asymptotic weight approached, an estimate of the mature weight (kg)

B = the rate of exponential decay of the initial growth rate, a measure of the decline in growth rate (kg), and

M = the age at which growth is maximum (days)

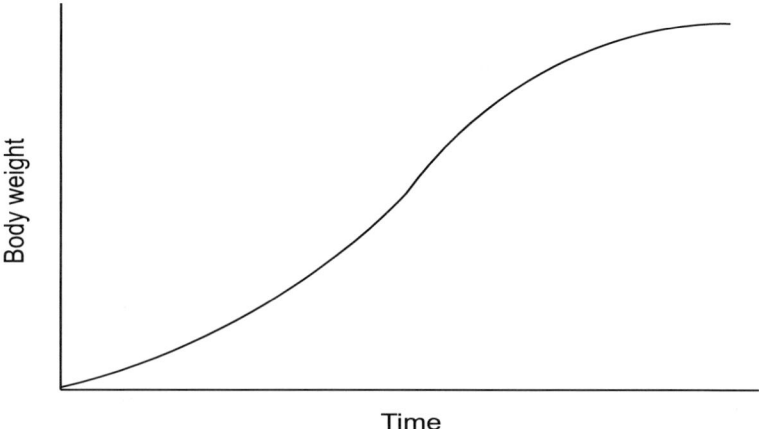

In all breeds except the Redbro, feeding presumed non-limiting rations gave earlier growth inflexions than feeding Label Rouge type rations. The fitted curve for the Redbro when fed Label Rouge type rations provided an over-estimate of early growth rate, and this explains why the rate of exponential decay in initial growth rate was high (Gompertz parameter B). Growth inflexion occurred at about day 38 in Ross 308 birds (commercial broilers)

Table 7. ESTIMATED GOMPERTZ B, M, C AND A FUNCTION PARAMETERS, FOR AS-HATCHED LIVE WEIGHT OF SEVERAL HYBRIDS AND TRADITIONAL BREEDS WHEN FED EITHER PRESUMED NON LIMITING RATIONS (NL), OR LABEL ROUGE TYPE RATIONS (LR) WHEN GROWN IN A CONTROLLED ENVIRONMENT

Breed	*Diet*	*Gompertz parameter*				
		B	*M*	*C*	*A*	P^a
(a) Fast						
Ross 308	NL	0.0438	37.8	6.010	0.011	99.8
Ross 308	LR	0.0314	54.8	6.785	0.028	99.8
Ross 508	NL	0.0411	38.9	6.106	-0.016	100.0
Ross 508	LR	0.0281	54.6	6.839	-0.035	99.8
(b) Moderate						
Redbro	NL	0.0306	44.4	5.505	-0.086	99.9
Redbro	LR	0.0727	31.5	3.161	-0.034	96.4
Master Gris	NL	0.0276	47.1	4.971	-0.104	99.8
Master Gris	LR	0.0291	50.1	4.478	-0.030	99.9
(c) Slow						
I657	NL	0.0276	49.2	4.275	-0.049	100.0
I657	LR	0.0203	65.5	4.950	-0.077	99.8
(d) Very slow						
Light Sussex	NL	0.0221	63.3	3.810	-0.029	99.4
Light Sussex	LR	0.0150	97.0	5.480	-0.037	99.0
Ixworth	NL	0.0237	56.8	3.899	-0.048	99.8
Ixworth	LR	0.0166	86.6	5.053	-0.039	100.0

Where P^a = percentage variance, and data are taken from a larger data set.

fed presumed non-limiting rations, whereas a Label Rouge type ration delayed the age at growth inflexion, to about day 55. Ross 508 birds were similar ages at growth inflexion to Ross 308 (Ross 508 birds have been selected for increased breast meat yield). ISA Master Gris birds were slightly faster growing when fed presumed non limiting rations than when fed Label Rouge type rations. Growth inflexion occurred at days 47 and 50, respectively. The ISA 657 birds had a growth inflexion at about day 49 when fed presumed non limiting rations, and a growth inflexion at about day 66 when fed Label Rouge type rations. ISA 657 are widely grown under the Label Rouge production system in France, where the minimum age at slaughter is 81 days. The largest differentials in age at growth inflexion between birds fed presumed non-limiting rations and birds fed Label Rouge type rations were for the traditional breeds. Growth inflexion in the Ixworth and the Light Sussex occurred at days 57 and 63 when fed presumed non limiting rations, but when fed Label Rouge type rations growth inflexion did not occur until days 87 and 97, respectively. This may be due to the digestive system developing more slowly in traditional breeds. Since feed intakes were lower when fed presumed non-limiting rations, compared with those of birds fed Label Rouge type rations, gut capacity was unlikely to have been a limiting factor. By comparison, gut capacity may have limited feed intake when fed Label Rouge type rations, and if intakes were lower than required for optimal growth at the given nutrient specification then this would have accounted for the delay in growth inflexion.

There are implications for eating quality associated with growth rate, as flavour tends to be enhanced after the growth inflexion (Moran, 1999 citing Touraille *et al.,* 1981 and Scholtyseek and Sailor, 1986). Thus, for table birds, where flavour is expected to be more intense than for broiler meat, growth inflexion should occur prior to the minimum age at slaughter. Based on the Gleadthorpe data shown in Figure 2 and Table 7, broiler hybrids may be used in free range table bird production as they have appropriate growth profiles, and growth inflexion occurs prior to slaughter at day 56. Ideally, females only would be used in free range table bird production as growth inflexion occurs earlier in females than in males and by using females only the meat may be highly flavoured.

There is perhaps more chance that flavour will be enhanced in ISA 657 birds fed Label Rouge type rations and grown to 81 days of age, as growth inflexion occurred well in advance of the minimum slaughter age. Traditional breeds such as the Ixworth would require a longer growing period than 81 days to meet required market live weights of between 2.0 and 2.5 kg, and from the fitted Gompertz curve it is estimated that birds would be at least 92 days of age when grown in a controlled environment (Figure 6). Growth inflexion was predicted to occur at about day 87, and so there may be some flavour enhancement when using an extended growing period, compared with growing birds to day 81 and selling them at a lighter weight; but this needs to be tested.

The heavier predicted mature live weights for modern broiler hybrids, compared with their ancestors, are due to a strong correlation between selection for rapid early growth and mature live weight (Prescott *et al.,*

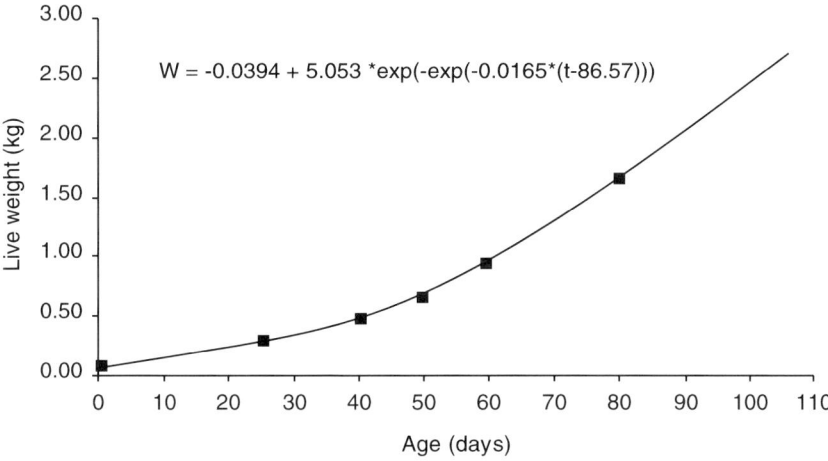

Figure 6. Gompertz prediction of live weight against age of Ixworth fed Label Rouge type rations (kg). Note that in this example the birds reached slaughter weight before the deceleration stage of the Gompertz growth curve. However, the equation was used in order to fit the specific data to a standard growth pattern.

1985, citing Merrit, 1974 and McCarthy, 1977). Prescott *et al.,* (1985), using the Gompertz function, predicted the mature live weight of male Ross broilers to be 6.43 kg, and 95% of mature live weight was achieved at day 153. In work at Gleadthorpe, the mature live weight of as-hatched Ross 308 broilers fed presumed non limiting rations was predicted to be 6.01 kg, and at the end of the growing period at day 81, 86% of mature live weight had been achieved. Using the fitted curve, it was estimated that 95% of mature live weight would have been achieved at about day 108 to 110, and 50% of mature live weight was achieved at about day 42 (Figure 7). This was much sooner than that reported by Prescott *et al.,* (1985), where 50% of mature live weight was achieved at day 60, and these authors cited the work of Brody (1945) where Brahama x Leghorn crosses reached 50% mature live weight at about 100 to 120 days of age. Strict comparisons are not possible because of differences in dietary specifications and photoperiods, but the reduction in time taken to grow is probably due to genetic and nutritional changes.

The comparison of findings between the work of Prescott *et al.,* (1985) and Gleadthorpe work, suggest that over the 15 year period there has been a considerable increase in the early growth rate of broilers. If similar increases in early growth rate are achieved over the next few years, then broiler hybrids of the future may have too fast a growth rate for using in free range table bird production.

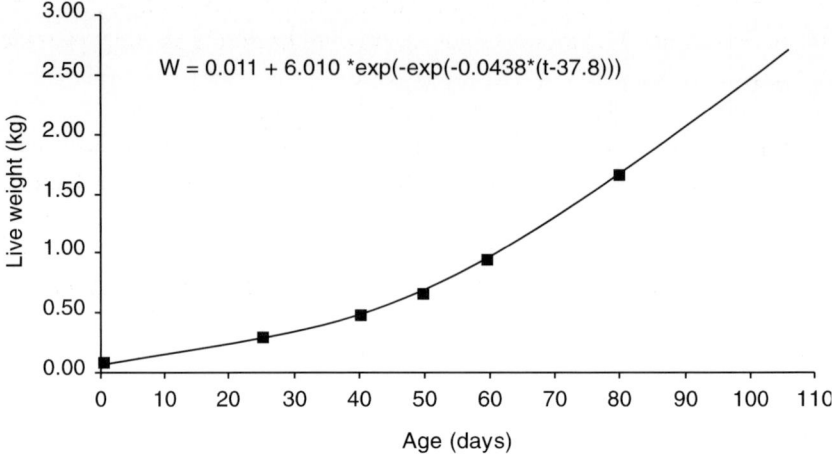

Figure 7. Gompertz fitted function to as-hatched Ross 308 live weight data (kg).

Taylor (1965) proposed the concept of metabolic age for mammals. He suggested a time-scaling parameter that was proportional to mature weight (A) raised to the power 0.27, and defined metabolic age (θ) as follows:

$$\theta = (t-3.5)/A^{0.27},$$

where, t = time in days and 3.5 is days from conception to the start of growth (i.e. at implantation); this component of the equation was described by the author as being the origin of the curve.

Laird (1966) proposed a time-scaling parameter for chickens that was proportional to mature weight (A) raised to the power 0.24 ($\pm$ 0.10). Thus, for chickens, it is suggested that metabolic age may be defined as:

$$\theta = (t+21)/A^{0.24}$$

where, t = time in days from hatch, and 21 is days from fertilisation to hatching (Gordon *et al.*, 2001).

As nutrient intake and growth are inextricably linked, and systems of production requiring slower growth are now developing, and meat quality including eating quality will be paramount to its success, attempts should be made to understand the biology and to compare breeds on a biological time scale equivalent. Metabolic age as defined by Laird (1966) may provide this basis.

There is evidence that meat flavour increases with age and this is discussed later. However, relevant to this discussion is the realisation that meat-eating qualities have been compared for breeds of different growth capacities either at a similar age, or at a similar live weight. If eating qualities are compared at a set live weight, such as typical market live weight, for breeds having different mature live weights, then the comparisons are not biologically consistent. At a typical market live weight, the breed having a lighter mature live weight may be more metabolically mature than a breed having a heavier mature live weight. The meat of the lighter breed may be highly flavoured because it is more metabolically mature than the heavier breed. Thus, it is suggested that experiments which aim to establish whether there are breed differences in meat flavour and texture should be designed so that comparisons are made at a similar metabolic age.

EFFECTS OF AMBIENT TEMPERATURE AND PHOTOPERIOD

Gross differences in the environment of intensively produced broilers and extensively produced table birds may affect feed intake and growth rate. Environmental temperature, ventilation rate, photoperiod and light intensity are controlled throughout broilers' entire growing period. By comparison, extensive production systems allow the birds' environment to be controlled for only up to half of the growing period, this being 28 days in the case of a free range system and 42 days in the case of traditional free range production systems. When the house popholes are open during the latter part of the growing period, so as to allow birds daily access to pasture, the house temperatures will fluctuate with ambient temperature; ventilation rate will exceed the minimum required for preventing the build up of noxious gases, and certain areas within the house may be draughty in windy weather. Whilst photoperiod and light intensity may be supplemented using artificial lighting, light quality will differ because natural daylight will filter through the popholes, and whereas artificial light sources have set peak wavelengths, the colour and intensity of daylight will vary depending on time of day and cloud cover. Environmental conditions on range will be dictated by seasonal changes in daylength and weather.

Increases in natural daylight hours should be avoided when growing birds to day 81 as this may stimulate early sexual maturity and this would be disastrous in as-hatched flocks as the ratio of males to females is high. Repeat mating will damage the females and, competition between males is likely to cause aggression. Lighting programmes should be used so that the daylength is held constant throughout the growing period. For spring flocks this will require the start of the light period to coincide with the earliest time of sunrise during the 81 day growing period, and the end of the light period to coincide with the latest sunset.

Seasonal changes in maximum and minimum air temperatures, and variability in air temperatures within a season, are shown in Figure 8 for ADAS Gleadthorpe during 1999. It can be seen that air temperatures vary widely over an 81 day period.

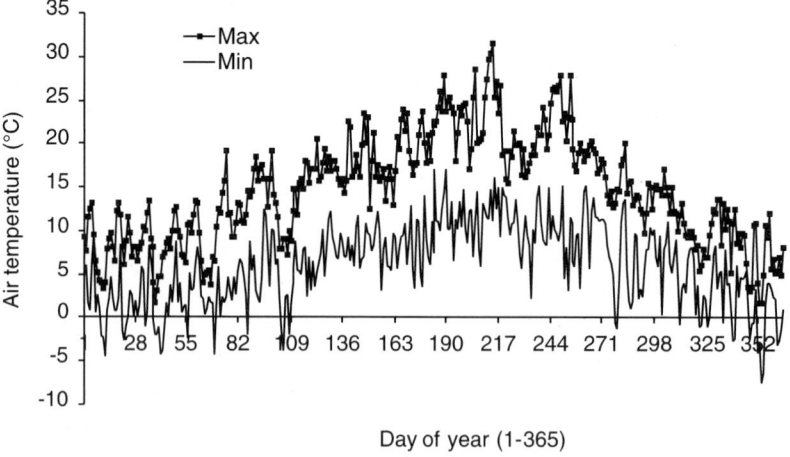

Figure 8. Maximum and minimum air temperatures (°C) at ADAS Gleadthorpe during 1999

Temperature influences growth largely by modulating feed intake. Feed intake is increased at temperatures below the birds' thermoneutral zone as the energy requirement for maintaining body temperature is increased, and as temperature increases feed intake is reduced (Cowan and Michie, 1978;

Hurwitz *et al.,* 1980; Charles *et al.,* 1981; Yahav *et al.,* 1996). Charles *et al.,* (1981) reported depressions in both feed intake and growth rate as temperature increased within the range of 15 to 27°C. The depression in live weight was not alleviated by increasing the nutrient density of the finisher diets (Gleadthorpe Poultry Booklet number 7). Multiple regression techniques were used to describe the response, and an economic model was developed (Wathes *et al.,* 1981), in which temperature could be optimised for the relevant conditions of live weight value and feed cost. Under UK market conditions, the economic optimum temperature, after the brooding stage, was usually between 20 and 23°C, depending on weight required at slaughter and sex of the birds. Other workers have reported that optimal post brooding temperatures fall within the range of 21 to 25°C (Meltzer, 1983a; Meltzer, 1983b).

In the UK, ambient temperatures are, for most of the year, not likely to be within the range 21 to 25°C, and in addition there will be daily fluctuations in ambient temperature. The maximum temperature may cause heat stress, or the minimum temperature may result in chilling. Based on a review by Sutcliffe *et al.,* (1987), it appears that broilers are able to tolerate consistent daily fluctuations in ambient temperature without a growth depression, provided that the mean daily temperature is close to the birds' zone of thermal comfort, and that the amplitude of temperature variation is not too large. Growth depression does occur when the amplitude of change varies, when the maximum daily temperature causes panting, or when extra feed energy is required to maintain body temperature. Later work by Yahav *et al.,* (1996) supports this view. Thus, it is expected that the growth rate of broiler hybrids grown to day 56 in a free range production system will be poorer than in birds housed in a controlled environment facility, and that the depression in growth rate will vary depending on the conditions experienced by the birds. This suggests that there will be a degree of uncertainty in free range production, which will make achieving precise market live weights more difficult than in broiler production. To the best of the authors' knowledge, optimum temperature regimes have not been established for slow growing hybrids and traditional breeds. Smaller birds are more susceptible to heat loss as their surface area to volume ratio is greater, and so in cool weather, feathering will be extremely important in reducing heat loss. The temperatures at which live weight is depressed are not known, and the effects of large daily fluctuations in hot, or cold temperatures on live weight gain are also unknown.

There will be a need for producers to adjust the metabolisable energy (ME) to crude protein ratio, and to adjust the mineral and vitamin concentrations in feeds at times of the year when extremes of temperature are experienced. Diets having a high energy to crude protein ratio will be required during cold weather because birds will be consuming feed energy in order to maintain body temperature. Conversely, in hot weather the ratio of metabolisable energy to crude protein may need to be reduced, as feed intake is reduced and reduced supply of essential amino acids could limit growth. At present, it is unlikely that feed energy and nutrient intakes will be optimal for birds grown in an 81-day free range production system, as little is known about the nutrient and energy requirements of slower growing birds.

Poultry can acclimatise to hot and cold weather (Meltzer, 1987). Birds acclimatising to hot weather develop enlarged combs and wattles, body fat tends to be reduced and feather cover is less than that of controls (van Kampen, 1981). Heat production is also lower than that of unacclimatised birds (Ariele *et al.,* 1980). The mechanisms involved in cold acclimatisation are both of an insulatory and metabolic nature. Non-shivering thermogenesis is the main mechanism in adult birds for maintaining body temperature (Meltzer, 1987 citing Freeman, 1983), and it is considered important in young birds (Ricklefs, 1985). However, feathering and the amount of insulation it provides is important. Differences in heat production between poorly and well feathered layers increase with decreasing ambient temperatures and reach a factor of 2 at about 0°C (van Kampen, 1981). Growing birds will not be fully feathered and heat loss is expected to be greater than for adult birds. Adequate intakes of sulphur-containing amino acids will be important in free range birds if feather cover is to be optimised prior to going out on range at day 42. This is because of the high sulphur-containing amino acid content of feathers (Larbier and Leclercq, 1992).

Photoperiod influences growth rate largely through effects on feed intake. When the photoperiod is 12 hours, or greater, feed intake is restricted to the light period and broilers rest during darkness (Savory, 1976; Gordon, 1994). Gut fill and passage time will ultimately limit feed intake during the shorter light period, and so feed intakes are lower in birds receiving a 12 or 16 hour photoperiod than in birds receiving a 20 hour photoperiod. Although feed intake occurs during very long scotoperiods it is at a lower rate than during the light period, and so feed intake is less than in birds receiving longer photoperiods (Gordon, 1995). Positive linear relationships were reported between feed intake and photoperiod, and between live weight and photoperiod, when photoperiods within the range of 8 to 20 hours were examined.

Producers may choose to use artificial lighting in housing facilities for free range birds, in order to extend the natural photoperiod, but it is not possible to apply shorter photoperiods than those occurring naturally. During the summer months, when temperatures may be so high as to cause heat stress, the houses will not be light-proof and birds will be active. In heat wave conditions it is common to feed broilers during the night when temperatures are cooler, and this necessitates the use of a near continuous photoperiod. Near continuous photoperiods are not permitted in some extensive production systems (e.g. free range according to Freedom Food Standards). So when producers are not able to use light to encourage birds to eat during the cooler night-time, feed intake and growth rate will be reduced in hot weather and, possibly, to a greater extent in extensively produced birds than in broilers.

Light quality is not expected to have large effects on growth rate, although comparisons of performance between birds grown in controlled environments but having natural light, *versus* birds grown using artificial light sources, are to the best of the authors' knowledge, not available.

Carcass growth and body composition

Carcass growth and body composition of broilers has been widely studied, mostly at a single age, or weight, which corresponds to slaughter age, or slaughter weight, and less frequently in terms of allometric growth and development. Allometric growth provides a better means for determining effects of age, sex, strain, nutrient intake and environment on body composition, but it is expensive both financially and experimentally, as birds are sacrificed throughout the growing period. To facilitate a structured approach to the review, carcass growth and body composition will be discussed under the sub-headings of age and sex of bird, strain, nutrient intake and environmental factors such as temperature and daylength. However, it is accepted that there are interactions between these factors.

EFFECTS OF AGE AND SEX OF BIRD

Growth of the body's muscles may not remain proportional to one another with change in age (Moran, 1999). Breast and thigh muscles differ in emphasis with development. Breast musculature is negligible at hatch, then extraordinary growth occurs for two weeks and then subsequently diminishes. Thigh muscles are well-developed at hatch, then their growth is moderate until two weeks and similar to breast thereafter (Iwamato *et al.,* 1993a). The embryonic development of thigh muscles enables pedal function and shivering thermogenesis at hatching (Ricklefs, 1985).

Muscle growth continues at the growth inflexion point and as skeletal development diminishes an increase in overall meat yield occurs (Lewis *et al.,* 1997; Moran, 1999). The breast musculature is a dominant contributor in this respect. Halverson and Jacobson (1970) reported that the meatiness of rapidly growing chickens was derived mainly from increases in weight, length and width of *M. pectoralis* and increases in weight and length, but not width of *M. supracoracideus.* Acar *et al.,* (1993) reported third order polynomial responses for breast muscle yield when expressed as a proportion of live weight against age. Although males had a higher breast muscle mass than females from two to 12 weeks of age, when expressed as a proportion of live weight there were no differences in breast meat yields between sexes. Other workers have reported similar breast meat yields at day 49 for males and female broilers (e.g. Orr *et al.,* 1984).

During the period of sexual maturation the thigh muscles of cockerels develop markedly under the influence of endogenous androgen (Iwamoto *et al.,* 1993 citing Ono *et al.,* 1982). More specifically it is the *M. iliotibialis lateralis pars postacetabularis* that is greatly influenced by endogenous androgen, as *M. pubo-ischio-femoralis* is only weakly affected by the presence of androgens (Vollmerhaus, 1992). In the domestic fowl, *M. iliotibialis lateralis pars postacetabularis* is large, and it is involved not only in the abductive function, but also in the action of stretching the hip joint and flexing the knee joint during terrestrial locomotion. Its rapid development during the pre-sexual maturing phase is presumably to assist the male in finding suitable mates, and in subsequent mating. In the females, muscle growth relative to body weight gain is very small during the pre-sexual maturing phase and body weight gain is due to the development of the reproductive tract and adipose tissue (Iwamato *et al.,* 1993a, citing

Iwamoto and Takahara, 1971 and Iwamoto *et al.,* 1975). Their findings suggest that there may be differences between sexes in thigh meat yield during the pre-sexual maturing phase and beyond.

Prescott *et al.,* (1985) examined the allometric growth of male broilers grown to 95% of their predicted mature live weight. Heart weight as a proportion of live weight declined slightly with age. The liver and the alimentary tract (empty) reached peak weights in relation to live weight very early in growth. At about day seven, the liver weight per unit live weight was twice as high as that of birds at day 49. After day 49, both showed little relative growth. Other workers have found that as chickens mature the digestive tract decreases in weight relative to body weight (Katanbaf *et al.,* 1989; Susbilla *et al.,* 1994), supporting the suggestion that the digestive tract plays a central role in determining the rate of growth of the very young animal (Crompton and Walters, 1979; Katanbaf *et al.,* 1989; Nir *et al.,* 1993).

Rose (1997) discussed allometric growth of body parts. He gave the following equation which related the weight of a body part to the weight of the whole body.

$$\log (y) = \log (a) + k \log (x)$$

where:

x = body weight (kg)
y = weight of body part (kg)
k = allometric growth ratio (slope of graph of log (weight of part) against log (whole body weight))
a = intercept of same graph

The allometric growth ratio (k) for breast meat in the chicken was given as 1.21, and so the rate of breast muscle growth was high relative to whole body growth. If genetic selection over the last 15 years has been successful in altering allometric growth, and more specifically the weight of breast muscle in relation to live weight during early life, then the allometric growth ratio for breast meat would differ between modern broiler hybrids and traditional breeds like the Ixworth.

EFFECTS OF BREED

Broiler production became a commercial enterprise in the USA in about 1930 (Warren, 1958). The first broilers were pure Barred Plymouth Rocks, but by 1942 about 97% of the commercial broilers were crossbreds resulting from the crossing of the Barred Plymouth Rock male on the Rhode Island Red female, or the New Hampshire female. By 1950, there was a shift to the use of the pure New Hampshire. Warren (1958) reported major disadvantages of the barred crossbred, these having poor feathering and dark coloured pin feathers which detracted from the appearance of the dressed bird. By comparison, the New Hampshires had better feathering and lighter coloured pin feathers. The trend towards a predominantly white bird indicated that the broiler of the then near future would be that colour. Meat yield, and in particular the desire to improve conformation, was addressed by using the "well-proportioned" Cornish Game as one parent of the crossbred broiler, although this caused some deterioration in growth rate, feathering and fertility.

Early studies on meat conformation tended to consider finish and edible cooked meat yields. This reflected the fact that birds were sold whole rather than as portions or fillets. Edible cooked meat yields were assessed by Dawson *et al.,* (1958) at 6 and 16 weeks of age in a commercial egg laying strain (White Leghorn), and three broiler strains (White Rock, M.S.U. Cornish Cross and Cornish x White Rock). At both 6 and 16 weeks of age, the percentage of cooked meat from White Leghorns was lowest and the percentage of bone was the highest. At 6 weeks of age, the M.S.U. Cornish Cross had the highest meat yields, but at 16 weeks of age the Cornish x White Rock had the highest yields. The percentage of cooked meat from the four groups increased from 49% at 6 weeks, to 51% at 16 weeks of age.

Over a period of 70 years, breeders have focused their efforts on improving finish and more specifically, breast meat yield. This is because of its extensive contribution to the total and economic value of poultry meat. Breast meat is now widely used by the catering industry and as an ingredient in ready meals.

Many comparisons of meat yields between early broilers and modern broilers of the time have been made. In earlier work, yields were reported for light and dark meat as opposed to breast meat yield and thigh and drumstick yield. An example, is that of Simpson and Goodwin (1979), who reported yield comparisons for Cornish Game hens and broilers. Light meat yields of the Cornish Game were lower on a weight basis than those of the broiler but, when expressed on a percentage basis, yields were similar. Dark meat yields were higher than white meat yields for both strains (approximately 55% dark meat yield, compared with approximately 43% white meat yield, as a percentage of dressed carcass weight).

Other workers have provided more details on carcass component yields for modern broilers of the time (e.g. Broadbent *et al.,* 1981). The authors reported a killing-out percentage of about 70% (eviscerated carcass weight plus giblets expressed as a proportion of live weight), breast meat yields of about 28% (excluding skin and bone), and leg meat yields of about 31% (including bone). Gizzard, heart and liver yields expressed as a proportion of the weight of the eviscerated carcass plus giblets were about 14%, 0.7% and 32%, respectively. They were not convinced that increases in broiler live weight gain had been accompanied by any major changes in the relative yields, either of the major body parts, or the components. However, even a 2% increase in breast meat yield may be important to the consumer, and several workers have reported increases in breast meat yield of this order. For example Orr *et al.,* (1984) reported strain differences in breast meat yield at day 49, the yields ranging from 22.2% to 24.2%.

The basis for improvements in breast meat yield is that the response to selection is not uniform over all tissues or components (Ricklefs, 1985). The author cited work by Ricklefs and Marks (unpublished data) who compared yields of control and selected lines of Japanese Quail and found that the only component that increased as a result of selection was pectoral muscle mass. Pectoral muscles of selected birds had a pronounced increase in exponential growth during the first week after hatching, whereas the growth rate of the leg muscles and the carcass (minus integument, muscle and viscera) increased less in response to selection. Interestingly, Ricklefs (1985) proposed that selection against leg muscle mass, whilst increasing the growth rate of other parts of the body, might tend to prevent genetic correlations and functional relationships among body components. Leg muscle developmental maturity was highly correlated with the Gompertz growth rate constant (0.85 correlation), and selection for increased breast meat yields in broilers had already been shown to alter the functional capacity of skeletal muscle (Ricklef, 1985 citing unpublished data of Kuenzel). Chicks selected for increased growth and breast meat yields were less able to maintain body temperature when exposed to a low ambient temperature (10°C) over a one hour period. The findings implied that shivering thermogenesis was impaired.

Selection for increased breast meat does not appear to be fully realised until the inflexion point in growth (e.g. Acar *et al.,* 1993). The authors measured *pectoralis major* development between two commercial sources of broilers known to differ in breast meat yield. Although the greatest proportion of breast muscle growth occurred in the first two weeks, strain differences did not maximise until the inflexion point, which occurred at seven to eight weeks of age.

Lewis *et al.,* (1997) compared breast meat yields at similar ages (day 83) and similar weights for Ross 1 broilers and ISA 657 (achieved by control feeding Ross 1), and also at similar ages (day 83) but different weights (achieved by *ad libitum* feeding Ross 1). When Ross 1 were control fed so as to achieve similar growth rates to ISA 657, the Ross 1 had negligible quantities of leaf and gizzard fat, but similar weights of breast, thigh, wing, skin, neck, tail and frame to ISA 657. On a percentage basis, Ross 1 had less thigh and frame, but more drumstick than ISA 657. Allowing Ross 1 to feed *ad libitum* resulted in all of their component parts being heavier than those of ISA 657. On a percentage basis, Ross 1 had higher breast meat yields, higher total meat yields, more leaf and gizzard fat, and more skin, but less wing than ISA 657. Greater wing flapping in ISA 657 was thought to account for the latter, as usage increases both bone mass and supporting muscle mass.

It seems that selection for an increase in the early growth rate of broiler breast muscles produces about 2% more breast meat at market live weight than in earlier hybrids. However, to achieve the higher breast meat yields the birds need to be fed *ad libitum* a high specification diet. This will not be the case when broiler hybrids are used in free range table bird production and, so breast meat yields are expected to be lower than for conventional broilers.

EFFECTS OF DIET SPECIFICATION AND FEEDING STRATEGIES

It is standard practice in the broiler industry to use formulation techniques for producing estimated optimal rations at least cost. Minimum nutrient requirements at a particular stage of growth for example, as published by the National Research Council (1994), provide the basis for formulation. Although least cost formulation techniques will be used for formulating diets for free range, or traditional free range meat birds, there are dietary specifications that the feed formulator will need to adhere to. The high minimum dietary cereal contents stipulated for free range and traditional free range meat birds will reduce the crude protein content of the ration, compared with broiler rations. Thus, lysine and sulphur-containing amino acids, these being the first and second limiting nutrients for growth, will be reduced in diets fed to birds in extensive production systems. Metabolisable energy values may be slightly lower than for broiler rations, but if wheat is the predominant cereal in rations for extensively produced table birds then the impact on feed energy value will be less than that on crude protein content. Differences between high specification broiler rations and Label Rouge type rations in crude protein, lysine and methionine contents, and the ratio of metabolisable energy value (MJ/kg) to lysine content (g/kg), are shown in Table 8. The information is taken from experiments at Gleadthorpe (MAFF-funded project OF0153).

Table 8. NUTRIENT CONTENTS OF BROILER RATIONS AND LABEL ROUGE TYPE RATIONS

Nutrient (g/kg)	Broiler rations			Label Rouge type rations		
	Starter	Grower	Finisher	Starter	Grower	Finisher
Crude protein	210	210	185	182	161	159
Lysine	13.2	12.9	12.2	10.3	8.6	7.6
Methionine	6.6	6.2	5.8	3.7	3.1	2.9
ME(MJ/kg) to lysine (g/kg) ratio	0.97	1.06	1.14	1.19	1.45	1.64

Source: MAFF-funded project OF0153

Summers and Leeson (1985) reported greater carcass protein yields in broilers fed a diet containing 220 g/kg protein, compared with birds fed a diet with a crude protein content of only 160 g/kg (similar crude protein content to that of Label Rouge type rations). Later work identified a reduction in carcass protein content at six weeks of age as the dietary crude protein content fell from 230 g/kg to 170 g/kg (Summers *et al.*, (1988). Whilst carcass protein contents were not recovered by supplementing the 170 g/kg crude protein diet with sulphur-containing amino acids, they were recovered by supplementing with both lysine and sulphur-containing amino acids.

There are implications for carcass conformation, breast meat yield and carcass fatness associated with feeding rations having a high metabolisable energy to lysine ratio. Almost 60 years ago, Fraps (1943) found that rations having a low energy to protein ratio caused less fat deposition than rations having a high energy to protein ratio. This has been confirmed more recently by other workers (Bartov *et al.,* 1974; Gleadthorpe Poultry Booklet number 7, 1980; Moran and Bilgili, 1989).

Work at ADAS Gleadthorpe (Gleadthorpe Poultry Booklet number 9, 1982) identified differences in fat deposition between males and females fed rations having increasing energy to crude protein ratios (range 532 to 783 KJ/kg per 1% crude protein and a crude protein range of 234 to 158 g/kg, respectively). As expected, fat deposition was lowest in both sexes when fed the narrowest energy to protein ratio. In the males, fat deposition in both the abdomen and skin was highest when fed the widest energy to protein ratio, and this was due to an over-consumption of energy relative to a lower growth rate. Unlike the linear response for the males, the fattest females were those fed rations within the middle range of energy to protein ratios.

Meat formation is impaired when the dietary supply of essential amino acids is insufficient, and lysine is of particular concern because of its very high proportions in muscle protein (Moran, 1999). The author cited work of Roth *et al.,* (1990), in which breast muscle amino acid contents were measured in male broilers aged between five and eight weeks of age, after having been fed either 21.0, 27.3, or 33.5 g crude protein/bird.day. The total lysine contents of breast muscle increased with increasing dietary crude protein intake (135, 190 and 223 mg total, respectively), but percentage muscle lysine contents were similar across all intakes (approximately 10.6%).

Bartov and Plavnik (1998) reported improvements in broiler breast meat yield when feeding diets having a moderate excess of crude protein and higher concentrations of lysine and total sulphur containing amino acids than recommended by the NRC. Carcass fatness was reduced, compared with broilers fed a diet containing the NRC recommended energy to crude protein ratio, and recommended concentrations of lysine and total sulphur containing amino acids. Other workers have found that the lysine requirement for maximum protein accretion is higher than that required for maximum weight (Sibbald and Wornetz, 1986).

Although sulphur amino acid inadequacies may be common in broiler production, their potential impact is less than for lysine, as muscle methionine contents are lower (Moran, 1994). Roth *et al.,* (1990 cited by Moran, 1999) reported increases in breast muscle methionine contents as dietary crude protein intakes increased from 21.0 to 27.3 g/day (35 mg total, compared with 53 mg total), but increasing the dietary crude protein intake to 33.5 g/day provided no further increase in breast muscle methionine content. Percentage breast muscle methionine content was approximately 2.8% across all methionine intakes.

A low expected impact for dietary methionine content on breast meat yield may not be valid in traditional free range production, as dietary methionine contents are typically half these of broiler rations. Furthermore, at a dietary crude protein content of 159 g/kg (this being typical of Label Rouge type rations) and a feed intake of 160 g/bird.day, or less, the daily crude protein intake would be at most 25.4 g/bird. This suggests that methionine intake, as well as lysine intake, in birds fed Label Rouge type rations may be less than required for optimal breast meat yield, although it will depend on genotype potential.

Sulphur-containing amino acids are the predominant amino acids in keratin (Wheeler and Latshaw, 1981), and good feathering will be important in insulating birds in extensive production systems. Larbier and Leclercq (1992), provided values for the essential amino acid contents of carcass and feathers of young chickens, and these are given in Table 9.

There have been some experiments that have compared meat yields of birds fed conventional broiler rations with meat yields of birds fed Label Rouge rations.

Lewis *et al.,* (1997) reported heavier breast, thigh, drumstick, wing, skin, bone and total meat yields for birds fed commercial broiler diets, compared with birds fed Label Rouge type diets. When expressing component weight as a percentage of the carcass weight, UK diets produced birds with larger

Table 9. ESSENTIAL AMINO ACID COMPOSITION OF THE CARCASS AND FEATHERS OF YOUNG CHICKENS[1] (g/100g PROTEIN)

	Carcass	*Feathers*
Lysine	6.5	1.6
Sulphur amino acids	4.0	7.9
Tryptophan	0.9	0.7
Threonine	4.2	4.6
Leucine	7.2	8.5
Isoleucine	4.3	6.4
Valine	4.7	8.9
Histidine	3.5	0.7
Arginine	6.8	7.3
Phenylalanine + tyrosine	7.0	7.4

[1]Age of birds not stated
Source: Larbier and Leclercq (1992)

thighs and smaller drumsticks, but breast meat yields were similar between both diets. Breast meat yields were 26.4% for birds fed commercial broiler diets and 26.2% for birds fed Label Rouge type rations.

In an experiment at Gleadthorpe breast meat yields were measured in a large number of commercially available hybrids and traditional breeds, when fed either presumed non limiting rations (with respect to commercial broiler hybrids widely used in the UK during 1999), or Label Rouge type rations. Breeds were sampled at day 59, provided that an as hatched live weight at day 56 of at least 2.0 kg had been achieved. Breeds not sampled at day 59 were sampled at day 87. Breast meat yields at day 59 are shown in Tables 10 and 11 and at day 87 in Tables 12 and 13. Breast meat yields tended to be lower in birds fed Label Rouge type rations, compared with birds fed presumed non-limiting rations. The greatest differences in breast meat yields due to diet were in the fast growing stock. This was not unexpected as the Label Rouge type diet was thought unlikely to provide the 1994 NRC recommended daily crude protein, lysine and methionine intakes. Slower growing breeds were expected to have had lower dietary requirements for growth and maintenance, and the balance between supply and demand was anticipated to be better than for faster growing birds when fed Label Rouge type rations.

Thus, broiler hybrids used in free range table bird production will have lower breast meat yields than conventional broilers, because the ration used in free range production will not meet the birds' nutrient requirements for optimal breast muscle growth.

Table 10. BREAST MEAT YIELDS AT DAY 59 IN FAST GROWING COMMERCIAL UK BROILERS (%)

Sex of bird	Treatment	Breast meat yield (%)
Males	Feed type:	
	• Presumed non limiting	30.1
	• Label Rouge type	23.1
		$p=0.12$
		sed±1.392
	Breed:	
	• Ross 308	25.3
	• Ross YA x PM3	26.6
	• Ross 508	27.9
		$p=0.13$
		sed±0.970
	Interaction	$p=0.16$
		sed±1.787
Females	Feed type:	
	• Presumed non limiting	32.3
	• Label Rouge type	26.1
		$p=0.19$
		sed±1.930
	Breed:	
	• Ross 308	28.5
	• Ross YA x PM3	29.0
	• Ross 508	30.1
		$p=0.23$
		sed±0.744
	Interaction	$p=0.23$
		sed±2.112

Source: MAFF-funded project OF0153

Table 11. BREAST MEAT YIELDS AT DAY 59 IN IMPORTED MODERATE-FAST GROWING COMMERCIAL HYBRIDS (%)

Sex of bird	Treatment	Breast meat yield (%)
Males	Feed type:	
	• Presumed non limiting	22.8
	• Label Rouge type	18.8
		p=0.14
		sed±0.915
	Breed:	
	• I757	22.3
	• I957	20.6
	• Redbro	19.5
		p=0.11
		sed±0.915
	Interaction	p=0.07
		sed±1.432
Females	Feed type:	
	• Presumed non limiting	24.1
	• Label Rouge type	21.6
		p=0.15
		sed±0.588
	Breed:	
	• I757	22.8
	• I957	24.1
	• Redbro	21.6
		p=NS
		sed±1.931
	Interaction	p=NS
		sed±2.306

Source: MAFF-funded project OF0153

Table 12. BREAST MEAT YIELDS AT DAY 87 IN IMPORTED SLOW-MODERATE GROWING COMMERCIAL HYBRIDS (%)

Sex of bird	Treatment	Breast meat yield (%)
Males	Feed type:	
	• Presumed non limiting	25.4
	• Label Rouge type	24.4
		$p<0.05$
		sed±0.073
	Breed:	
	• I457	25.5
	• I657	25.1
	• Gris Barre	24.0
		$p=0.17$
		sed±0.656
	Interaction	$p=NS$
		sed±0.761
Females	Feed type:	
	• Presumed non limiting	27.0
	• Label Rouge type	26.9
		$p=NS$
		sed±0.696
	Breed:	
	• I457	27.9
	• I657	26.8
	• Gris Barre	26.7
		$p=NS$
		sed±1.027
	Interaction	$p=NS$
		sed±1.374

Source: MAFF-funded project OF0153

Table 13. BREAST MEAT YIELDS AT DAY 87 IN TRADITIONAL UK BREEDS (%)

Sex of bird	Treatment	Breast meat yield (%)
Males	Feed type:	
	• Presumed non limiting	22.8
	• Label Rouge type	21.3
		p=0.09
		sed±0.220
	Breed:	
	• Light Sussex	21.0
	• White Sussex	20.5
	• Ixworth	24.7
		p<0.01
		sed±0.663
	Interaction	p=NS
		sed±0.796
Females	Feed type:	
	• Presumed non limiting	24.0
	• Label Rouge type	21.9
		p=0.09
		sed±0.807
	Breed:	
	• Light Sussex	22.0
	• White Sussex	22.4
	• Ixworth	24.4
		p=0.19
		sed±0.646
	Interaction	p=NS
		sed±1.134

Source: MAFF-funded project OF0153

MacLeod *et al.*, (1998) addressed the question of the possibility of breed differences in response to amino acids due to differences in growth rate. They compared two modern broiler strains with a relaxed selection line with the growth characteristics of 40 years ago. Lysine requirement as a proportion of feed was found to be similar, and this was attributed to feed intake keeping pace with the improvements in growth rate.

Walker *et al.,* (1995) applied the Gompertz function to live weight data and selected carcass component data for broilers fed starter and finisher rations having either high, or low nutrient concentrations, and for comparison another group of birds was fed a commercial starter ration, followed by a commercial finisher ration. The crude protein concentrations of the six rations are given in Table 14. Changeover from starter to finisher rations occurred at day 21.

Table 14. DIETARY CRUDE PROTEIN CONCENTRATIONS (g/kg)

Calculated	*Starter*			*Finisher*		
analyses (g/kg)	*High (H)*	*Low (L)*	*Commercial (C)*	*High (H)*	*Low (L)*	*Commercial (C)*
Crude protein	248	199	223	220	183	201

Note: the low nutrient starter and finisher rations had higher crude protein concentrations than typical Label Rouge rations.

This work showed that the effects of nutrient intake on whole body growth rate were not always mirrored by similar changes in breast muscle and thigh muscle growth rate (Table 15). As expected, birds fed a high nutrient starter ration, followed by a high nutrient finisher ration were youngest at live weight growth inflexion, whereas birds fed a low nutrient starter ration, followed by a low nutrient finisher ration were oldest at live weight growth inflexion. The transition from a high nutrient starter ration to a low nutrient finisher ration delayed live weight growth inflexion, and it was marginally later than in birds fed a low nutrient starter ration, followed by a high nutrient finisher ration. Live weight growth inflexion occurred at about day 40 in birds fed commercial rations. Growth inflexions of breast and thigh muscles were earliest in birds fed a high nutrient starter ration, followed by a high nutrient finisher ration, and latest in birds fed a high nutrient starter ration, followed by a low nutrient finisher ration. Birds fed a low nutrient starter ration, followed by a high nutrient finisher ration had earlier growth inflexions for breast muscles, and thigh muscles, than birds fed a low nutrient starter ration, followed by a low nutrient finisher ration. Growth inflexions of breast muscles and thigh muscles occurred at about day 52 in birds fed commercial rations.

Table 15. ESTIMATES OF AGE AT GROWTH INFLEXION (DAYS) FROM
GOMPERTZ EQUATION FOR LIVE WEIGHT AND SELECTED CARCASS
COMPONENT DATA

	H-H	H-L	*Diet* C	L-H	L-L
Live weight	29	40	40	38	42
Breast muscle	38	65	52	42	47
Thigh/leg muscle	34	53	46	42	45

It seems that dietary crude protein concentrations have a greater impact on muscle growth rate than on whole body growth rate, and that breast muscle growth is affected more than leg muscle growth rate.

To relate the findings of Walker *et al.,* (1995) to free range table bird production, the diet combination L-L should be used, although it is accepted that the crude protein contents of the rations were somewhat higher than for Label Rouge type rations. It would seem that, when using broiler hybrids in free range table bird production, breast meat yield at the minimum slaughter age of day 56 may be about optimal and much longer growing periods would not be expected to increase breast meat yield. This is because maximal growth rates of the whole bird and of breast muscle are expected to have been reached at day 56.

EFFECTS OF AMBIENT TEMPERATURE AND PHOTOPERIOD

The effects of temperature on feed intake and growth rate in broilers have already been discussed; but to summarise, feed intake and growth rate are depressed as temperature increases within the range of 15 to 27°C. In cold ambient temperatures (7.5°C) feed consumption is increased, and both meat and fat deposition are decreased (Moran, 1999, citing the work of Smith and Teeter, 1987). In extensive systems of poultry meat production, birds will experience a much wider range of mean daily temperatures than broilers, and the extremes of daily temperature will also be greater than in controlled environments.

Breast meat usually suffers more than other parts of the carcass when body weight is not fully realised (Howlinder and Rose, 1989; Smith 1993).

Work at Gleadthorpe during the 1980s found that breast meat yields in males were reduced from 23.3% to 21.7% as temperature increased from 20 to 27°C (Gleadthorpe Poultry Booklet number 10, 1983). In contrast, females showed a curvilinear response to temperature, where breast meat yields were lower at 20°C and 27°C, than between 21 and 24°C. The optimum ratio of breast to thigh and drumstick meat was at 19.7°C in males, and at 21.4°C in females. When meat deposition was highest, this coincided with the maximum meat to bone ratio. Fat deposition increased in males as temperature increased up to 21°C, plateaued between 21 and 24°C, followed by a decline at temperatures above 24°C. In females, higher temperatures tended to reduce abdominal fat contents. Abdominal fat contents were higher in females than males. Fat deposition in the skin, when fed a high nutrient density diet, was highest in males at about 24°C, and highest in females at about 19°C to 21°C. The deposition of skin fat alters the appearance of the dressed bird so that it looks creamy, and skin fat bastes the bird during roasting. In extensively produced table birds where daily mean ambient temperatures are unlikely to be as high as 19°C to 24°C, skin fat may be less than needed for basting during roasting, although the effect of low temperatures on skin fatness may be negated by feeding a diet having a wide energy to crude protein ratio.

The need to alter dietary energy to protein ratios for free range birds so as to take into account seasonal differences in ambient temperature and feed intake has already been discussed. Marginal supplies of protein and essential amino acids are particularly influential in increasing body fatness (Lipstein *et al.,* 1975; Halvorson *et al.,* 1991; Cahaner *et al.,* 1995). Likewise, feather cover influences the birds' ability to either conserve or dissipate heat, which in turn affects fatness (Ajang *et al.,* 1993).

Yahav *et al.,* (1996) reported a progressive decline in broiler feed intake and weight gain as temperature increased within the range 18 to 35°C. Breast meat yield at day 56 decreased steadily with increasing temperature, whereas abdominal fat contents increased with increasing temperature. Diurnally cycling the temperature from 10°C to 30°C produced similar breast meat yields and abdominal fat pad yields to those of birds housed at a constant temperature of 20°C. In a further study, Yahov *et al.,* (1996) found that a diurnally cycling regime of 15 to 35°C reduced live weight and breast meat yields at day 56, whereas abdominal fat pad yields were increased, compared with birds housed at a constant temperature of 25°C.

Other authors have reported lower breast meat yields at day 54 when using a diurnally cycling temperature regime of 21 to 31°C (Sonaiya *et al.,* 1990).

If a diurnally cycling temperature regime reduces live weight gain then breast meat yield is expected to be reduced and carcass fatness increased. In the UK, birds grown in extensive production systems are likely to experience diurnal fluctuations in temperature, and daily increases, or decreases in mean daily temperature that are likely to affect live weight gain, breast meat yield and carcass fatness. Furthermore, males and females may respond differently. There will be interactive effects of strain, sex, temperature and nutrient intake on live weight gain, breast meat yield and carcass fatness, but it is not possible to quantify them using current knowledge. From a practical perspective this means that it will be difficult, if not impossible, to optimise nutrient intake in extensively produced table birds. It will also be difficult to minimise variation in carcass quality, so that product specification will be inconsistent.

Reducing the photoperiod within the range of 20 to eight hours per day reduced live weight gain (Gordon, 1995). As discussed previously, when live weight gain is reduced breast meat yield tends to suffer more than dark meat yield. Renden *et al.,* (1993) reported lower breast meat yields at day 49 in male broilers receiving either a 14, or 16 h light day, compared with birds receiving a 23 h light day. Thus, if producers do not use artificial lighting to supplement very short natural photoperiods, or if there are seasonal influences on growth regardless of artificial lighting, then breast meat yields and carcass fatness may be affected.

EFFECTS OF EXERCISE

Exercise has been shown to affect muscle growth in broilers (Sandusky and Heath, 1988a). Barriers of differing heights were placed in pens and as the birds grew the barriers were increased in height. The barriers produced changes in exercise patterns, and the weights, lengths, widths and depths of individual muscles were altered. Muscle dimensions (including breast, drumstick and thigh muscles) changed with barrier height and this was thought to be due to differences in the amount of work required to cross them. There were interactive effects of barrier height and sex on muscle dimensions and these were associated with crossing strategy. The males were heavier at a given age and they adopted a different crossing strategy to the females, and males crossed the barrier less frequently.

In extensive production systems where birds may be encouraged to be more active than conventional broilers, exercise may influence muscle dimensional relationships and carcass conformation, but the overall effect is unknown. The inferences from the work above are that the length of breast, drumstick and thigh muscles of extensively produced birds will be longer than those of broilers. The possible effects of exercise on meat texture are discussed later.

Muscle structure, muscle growth, muscle fibre type and changes in fibre type with age

In order to describe the effects of experimental or practice growing treatments on lean meat quality, some account of muscle growth is needed. The following is a summary.

Muscle is a living organ that is converted to meat *post mortem*. Muscles vary in function, and in order to meet functional demands they vary in composition. The composition of muscle (including nervous, vascular, adipose and connective tissues) and fibre size will influence meat texture. It is relevant to this chapter to consider muscle structure and growth, and to report differences in muscle fibre type proportions that occur with age, and differences that occur between meat-type strains and egg-type strains, as there are implications for meat texture.

MUSCLE STRUCTURE

Skeletal muscle is made-up of elongated multinucleated cells, or myofibres specialised for contraction (Wicks, 1999). Each myofibre is approximately 100 μm thick and contains many myofibrils, each about 1 to 2 μm thick. In striated, or skeletal muscle tissue, myofibrils display a pattern of alternating light and dark bands. The striations arise as a result of the packing arrangement of the filament systems in the sarcomere, the fundamental contractile unit of striated muscle. Each sarcomere consists of one set of thick (myosin filaments) and two sets of thin (actin) filaments, and during the contraction of the muscle the thin filaments are pulled over the thick filaments so that each sarcomere shortens and generates force (Goldspink and Yang, 1999 citing Huxley, 1969). The force to produce contraction is generated by the myosin cross-bridge, each cross-bridge interacts with a thin filament to pull it towards the centre of the sarcomere. The cross-bridge then detaches from the thin filament and has to be reprised by ATP before it can go through another cycle of force generation. *Post mortem* ATP keeps muscles extensible and there is little change in muscle length, this is known as the delay phase prior to the onset of rigor (Warris *et al.,* 1999). During rigor the extent of muscle contraction is important as it determines meat tenderness (Lyon and Buhr, 1999, citing Locker, 1960).

Different types of muscle fibres have different myosin cross-bridges and this allows differences in the speed of contraction and the differences in muscle fatiguability (Goldspink and Yang, 1999 citing Goldspink, 1985). If muscles are composed of differing fibres then *post mortem* fibre shortening within a muscle may not be uniform and this may adversely affect meat tenderness. The *post mortem* conversion of muscle to meat, and the relevances of ATP usage and muscle fibre type composition to meat texture are discussed later in the review.

MUSCLE GROWTH

Differences in the rate of muscle growth between breeds, and the physiology of muscle growth have been of interest for many years. Early work found that chicken embryos of heavy breeds proliferated cells faster than embryos of light breeds (Smith, 1963 citing the work of Blunn and Gregory, 1935). Differences in cells numbers between breeds were apparent as early as 72 h following the beginning of incubation. Smith's own work confirmed this. He determined that increases in muscle cell numbers in meat-type strains occurred before hatching, and that post hatching, growth was primarily due to increases in muscle cell size. Reciprocal crossing of a meat-type strain and an egg-type strain resulted in muscle cell numbers at hatching, and cell size at 10 weeks of age, being intermediate between the sire and dam. A higher number of muscle cells at hatching in meat-type strains means that mature body weights will be greater than that of more egg-type strains.

Muscle fibres increase in diameter throughout the growth period (Remignon *et al.,* 1994), and this is due to marked increases in myofibrilar content. Each myofibril reaches a certain size, after which it splits into two or more daughter myofibrils (Goldspink and Yang, 1999). Increases in muscle fibre length are due to the serial addition of new sarcomeres onto the ends of existing myofibrils. In this way, sarcomere lengths are constantly adjusted back to the optimum required for force production. Fibre diameter is important in determining meat texture and, in particular firmness and stringiness. Large fibre diameters may increase meat firmness, whereas smaller diameter, closely packed fibres may feel stringy in the mouth. The effects of production factors on cooked poultry meat textures are discussed later in the review.

Satellite cells (potential muscle myoblasts), which are found adjacent to existing skeletal muscle fibres, fuse with these fibres, and their nuclei direct the synthesis of new protein during the maturation of muscle. These events are controlled by specific growth factors that are produced locally by the satellite cells and other cells in muscle. There is a higher proportion of active satellite cells in young growing animals than in older animals (Goldspink and Yang, 1999 citing Snow, 1977). Satellite cells account for about 30% of the nuclei in muscle fibres in neonates and about 2-4% in adults' fibres. In general, oxidative muscles have a much higher density of satellite cells than glycolytic muscles (Goldspink and Yang, 1999). From a production perspective, a better understanding of the mechanisms regulating satellite cell activity may lead to the development of practices that increase the deposition and efficiency of lean muscle accretion, or perhaps improve the nutrient composition of meat products (MacFarland, 1999).

The rate of accumulation of muscle protein, the key element in muscle growth, is the result of the balance between the rates of protein synthesis and of protein degradation (Kang *et al.,* 1985). The authors cited the work of Maruyama *et al.,* (1978) where differences in growth rate at two weeks of age between meat-type chicks and layer-type chicks were due to slightly higher rates of muscle protein synthesis and markedly lower degradation rates. Their own work, using broiler chicks, identified that at one week of age as much as 68% of the protein synthesised in the breast muscle is retained, compared with only about 40% in Leghorn chicks (Kang *et al.,* 1985, citing a comparison of their own work with data taken from MacDonald and Swick, 1981). In a review of muscle growth and development,

Dransfield and Sosnicki (1999) reported that slow growing birds have a higher ratio of proteolytic enzymes to inhibitors than faster growing birds, thus supporting the view that increased growth and muscle mass in modern lines of broilers may be governed by reduced rates of protein catabolism.

Lysine deficiency reduces the amount of breast muscle protein gained each day and the amount of protein synthesised, whereas only slight changes in the absolute rates of protein breakdown were identified (Tesseraud *et al.,* 1996). *M. pectoralis major* was the most sensitive of muscles studied (other muscles studied were *M. anterior latissimus dorsi*, a wing muscle, and *M. sartorius*, a leg muscle). The authors suggested that leg and wing muscles may be protected from diet-induced atrophy through exercise, whereas breast muscles serve little functional purpose in broilers. Thus, when the dietary supply of nitrogen and essential amino acids are less than optimal, as in Label Rouge diets, a reduction in breast yield will be unavoidable.

MUSCLE FIBRE TYPE, CHANGES WITH AGE AND BREED DIFFERENCES

Early investigations classified muscle fibres into two arbitrary classes, red and white. Red fibres are generally smaller in diameter and contain more mitochondria than white fibres, and red fibres contain the pigmented protein myoglobin. Red fibres have more blood supply and contain a higher concentration of lipid than white fibres.

It is now widely accepted that the above classification is incomplete due to the presence in each of the classes of multiple myosin isoforms (reviewed by Wick, 1999). In this review, muscle fibre type classifications as provided by Remignon *et al.,* (1994) will be used. They described five fibre types in birds, of these three are slow twitch (types 1, 3A and 3B) and two are fast twitch (2A and 2B). Type 3 fibres are specific to birds and have the characteristics of being multi-innervated. The other types (1, 2A and 2B) are common to both mammals and birds. Of the fast twitch muscle fibres, type 2B (fast glycolytic) are adapted for high power output over a short period, whereas type 2A (fast oxidative glycolytic) are adapted for high power output over a longer period (Goldspink and Yang, 1999). Both of the type 2 fibres have forms of myosin and other contractile proteins that produce a fast cross-bridge cycle time and develop force rapidly. Type 1 fibres (slow oxidative) are adapted for slow repetitive movements and the fibres have a form of myosin which hydrolyses ATP slowly, thus resulting in a slow cross-bridge cycle. Other types of slow twitch fibres (type 3, and known as slow tonic) are found in muscles that are contracted for most of the time, such as the *M. anterior latissimus dorsi* whose function is to hold the wing back against the body.

Muscle fibre types differ phenotypically in that they not only express different subsets of myofibrillar isoform genes with different ATPase activities, but also different types and levels of metabolic enzymes. This means that skeletal muscles have an inherent ability to adapt to mechanical signals by inducing or repressing the transcription of different isoform genes and altering the expression of different subsets of genes. As there are several myosin isoform genes this means that muscle fibres possess the ability to alter their contractile properties during development, and in response to levels of activity (Goldspink and Yang, 1999 citing work by Moore and Goldspink 1985 and Goldspink *et al.,* 1992). This involves both qualitative and quantitative changes in gene expression which result in an alteration in the cross-sectional area of a given fibre type.

Iwamato *et al.,* (1993b) examined breed differences in growth of the *M. pubo-ischio femoralis pars medialis* (PIF). The PIF muscle is important in maintaining posture and supporting body weight (Iwamato *et al.,* 1993b citing Raikov, 1985). It is made up of two parts, the *pars medialis* and the *pars lateralis*, and although the former contains predominantly type 1 fibres, while the latter is almost completely composed of type 2A fibres, the functions are thought to be similar. PIF weights increased steadily with increasing body weight, but in some breeds, notably the White Cornish (broiler type), measured muscle weights and areas were smaller than the values calculated using regression equations, whereas in lighter breeds (Shamo, a Japanese game fowl), and the New Hampshire, the opposite was found. Histological examinations revealed that the development of type 1 fibres was observed only in the cranial part of the PIF muscle in the White Cornish. Surprisingly, this suggests that the caudal part, the main part of the PIF muscle involved in postural function, does not develop adequately in this breed.

A later study by Iwamato *et al.,* (1993a) examined muscle growth and fibre development in two skeletal muscles having differing growth rates and fibre type populations (PIF composed of type 1 and type 2A fibres, and *M. iliotibialis lateralis pars postacetabularis*, ITL, composed of type 2B fibres). The roles of the ITL muscle are to assist in stretching the hip joint and flexing the knee joint during walking, and its growth is stimulated by androgens (Iwamoto *et al.,* 1993a citing Vollmerhaus, 1992). Based on their findings Iwamato *et al.,* (1993a) suggested a four phase growth process for skeletal muscles. An early stage having two phases, namely a post hatching phase (from hatching to two weeks), followed by a marked growth phase (up to 15 weeks of age). The later stage also contained two phases, a sexual prematuration phase (from 15 to 20 weeks of age), and a sexual maturation phase (until 35 weeks of age). During the latter stage, skeletal muscle attains its full growth.

Growth in the early stage is characterised by increases in muscle fibre length, together thickness and diameter. In the early post hatching phase, regional differences in muscle growth are very large. Muscles composed mostly of type 2B fibres, such as the breast muscles (Iwamato *et al.,* 1993a citing Iwamato *et al.,* 1984), and the ITL muscle, are relatively light and contain many small fibres at the time of hatching, whereas muscles composed of fibre types 1 and type 2A, such as the PIF muscle, are larger at hatching. During the post hatch phase (day old to 2 weeks of age) there are large increases in the diameter of type 2B fibres (breast muscles are composed almost entirely of type 2B fibres) and only small increases in type 1 and 2A fibres (as found in the PIF muscle). In the marked growth phase (two to 15 weeks of age) large relatively constant growth rates are observed across individual muscles, although the rates decrease with advancing age. In the later stage, when increases in muscle length and fibre length are minimal, growth results from an increase in muscle thickness due to increases in fibre diameter. During the sexual prematuration phase (15 to 20 weeks of age) the proportional muscle weight maintains a steady increase in both sexes. However, in the sexual maturation phase (20 to 35 weeks of age) there are large differences in muscle growth between the sexes. Muscle development in cockerels under the influence of androgen is observed particularly in the hind limb muscles with the development of fibres, both in quantity and size (Iwamato *et al.,* 1993a citing Ono *et al.,* 1982, 1983). Conversely, in females, muscle growth rate is very low in relation to body weight gain, which abruptly increases because of the development of the female genital system, and the deposition of adipose tissue (Iwamato *et al.,* 1993a citing Iwamato and Takahara, 1971; Iwamato *et al.,* 1975).

At hatching all ITL muscles fibres were type 2A, but at only one week of age type 2A fibres were seen to differentiate into type 2B (Iwamato *et al.,* 1993a). The proportion of type 1 transitional fibres in the PIF muscle were high until two weeks after hatching, and thereafter, the proportion fell until the adult level was reached at 10 weeks of age.

Based on the four phases of muscle growth proposed by Iwamato *et al.,* (1993a), there are likely to be increases in muscle fibre diameter towards the end of the 12 week growing period for traditional free range meat birds, and this may influence cooked meat texture. Also, depending on the breed and the influence of the light environment, males may be entering the presexual maturation phase of muscle growth, during which time the number and diameter of thigh muscle fibres both increase. Variability in cooked thigh meat texture between birds may become apparent to the consumer over time.

According to Aberele *et al.,* (1979) fibre size and fibre type composition of skeletal muscles differ between meat-type and egg-type chickens. Broilers have larger muscle fibre diameters than egg-type strains, and broilers have a higher proportion of white fibres and a lower proportion of red fibres than egg-type strains (identified using succinic dehydrogenase staining). The authors suggested that broiler thigh meat may be whiter than layer thigh meat, because the muscles of broilers are more anaerobic.

However, other workers have failed to identify differences in fibre type proportion between cockerels divergently selected for fast, or slow growth (Remignon *et al.,* 1994). Fibre type proportions of *M. pectoralis major* (PM), *M. anterior latissimus dorsi* (ALD) and *M. sartorius* (SART) were studied at 11 and 55 weeks of age. In the ALD muscle, only slow tonic avian myofibres (type 3) were found, whereas in the PM muscle only fast twitch fibres were found (type 2) and they were predominantly type 2B (greater than 99%). In the SART muscle, two distinct parts were identified. The superficial part contained all fast twitch fibres (type 2), while the deepest part contained both fast twitch (type 2) and slow twitch fibres (type 1). At 11 weeks of age, the ALD muscle of the fast growing line contained a small proportion of type 2A fibres and fewer type 3B fibres than in the slow growing line, but at 55 weeks of age, there were no type 2A fibres in the ALD muscle of either line, and the proportions of type 3A and 3B fibres were similar between lines. There were no differences in fibre type proportions in the PM or SART muscles between fast and slow growing lines at either age. At 11 and 55 weeks of age, the PM muscle contained greater than 99% type 2B fibres. In the superficial area of the SART muscle there was a decrease in the percentage of type 2A fibres and an increase in the percentage of type 2B fibres with age. In the deep part of the muscle there was an increase in the percentages of types 1 and 2B fibres, and a decrease in the percentage of type 2A fibres.

Remignon *et al.,* (1994) reported that the fast growing line had larger mean cross-sectional fibre diameters for all fibre types, and for all muscles except in the deep part of the SART muscle where no differences were found for types 1 and 2B fibres at 55 weeks of age. Fibre size differences between the two lines were more marked at the younger age in all muscles. The cross-sectional area of type 2B fibres in the PM muscle at 11 weeks of age were 1.9 times greater in the fast growing line than in the slow growing line, but the size ratio was reduced to 1.3 at 55 weeks of age. In all muscles, the relative gain in mean cross-sectional area between 11 and 55 weeks of age was greater in the slow growing line than in the fast growing line.

Total muscle fibre numbers were estimated for the ALD muscle at 11 and 55 weeks of age, but not for PM and SART muscles. The fast growing line had about 25% more fibres in the ALD muscle than the slow growing line. Fibre numbers did not change with age. Interestingly, when comparing birds at similar live weights and PM muscle weights (i.e. comparing the fast growing line at 11 weeks of age with the slow growing line at 55 weeks of age), the younger birds from the fast growing line had smaller fibres. This suggests that there were more fibres in the PM muscle of the fast growing line, than in the PM of the slow growing line. There were similar findings for the SART muscle. They concluded that muscle fibre type composition was related to function rather than growth rate, whereas muscle maturation rate was affected by selection for growth rate.

It appears that breed differences in body weight and muscle weight are due to differences in cell proliferation during the embryonic phase, faster growing strains having a higher number of muscle cells at hatching; differences in muscle fibre diameter; and when measured prior to maturity, differences in metabolic age will also be relevant.

Extensive birds potentially receive opportunity for exerc ise or an examination of the effects of exercise on meat quality follows.

Exercise has been shown to affect fibre type proportions in leg muscles undergoing increased contractile work. Brackenbury and Williamson (1989) using White Leghorns, identified increases in the oxidative capacity of *M. illiotibialis lateralis caudalis* following repeated periods of treadmill training. This muscle is composed of type 2 fibres, and before exercise about 40% of these stained intensely for succinic dehydrogenase (red fibres). After six and 15 weeks of treadmill training the proportion increased to 50 and 60%, respectively.

Other workers have reported differences in muscle fibre type proportions between extensively reared ducks and intensively reared ducks (Dransfield and Sosnicki 1999 citing the work of Pingel and Knust, 1993). Ducks kept on pasture had more red fibres, and smaller diameter red and white fibres than those kept intensively. Dransfield and Sosnicki (1999) suggested that a higher proportion of smaller diameter muscle fibres may produce tougher, denser meat.

The role of exercise in influencing muscle fibre type composition and meat texture has not received a great deal of attention. This is because activity in broilers has traditionally been discouraged by using dim lighting, and in any case the consumers' preference for broiler breast meat has meant that broiler thigh meat is a low value product. However, exercise, or more specifically the encouragement of pasture usage in free range and traditional free range chickens, is something that UK consumers often cite as important. If pasture design, pasture management and bird management techniques are developed so that a large part of the birds' day is spent active (walking, running, wing flapping, scratching, dustbathing and perching), then activity may influence muscle size, dimensional relationships and fibre type proportions. Furthermore, extensively produced table birds have a higher ratio of dark to white meat than broilers, and so the value of the dark meat to the consumer will be greater than for broilers. This means that it will be important for producers to optimise both white and dark meat quality, and in order to do this a better understanding of the effects of exercise on muscle and associated tissue development will be needed.

It seems that intensively produced broilers fed high specification diets will have muscle fibres that are large in cross-sectional area at slaughter age. By comparison, when broiler hybrids are used in free range table bird production the cross-sectional areas of fibres may be smaller and, this may produce denser meat textures. Cross-sectional areas of thigh muscle fibres may be rapidly increasing in male traditional free range birds at slaughter age and, this may produce firmer and denser meat than in females.

The conversion of muscle to meat and the effects of production techniques on cooked meat texture are discussed later in the text.

References

References for Chapters 4, 5 and 6 on meat quality follow Chapter 6.

5

CARCASS DAMAGE, COLOUR AND NUTRIENT COMPOSITION

Carcass damage

Carcass damage that occurs prior to processing includes hock burn, breast blisters and bruising. The effects of age, breed, diet specification, ambient temperature and photoperiod on the development and incidences of hock burn, breast blisters and bruising are discussed.

EFFECTS OF AGE, BREED AND DIET SPECIFICATION

Brownish-black lesions which occur on the hock, breast and feet of broilers have been described collectively as contact dermatitis (Tucker and Walker, 1992 citing McIlroy *et al.,* 1987). Its incidence in intensive broilers in the UK has probably steadily declined in recent years, due to improved litter conditions. These have followed changes such as the use of enzymes in feed.

Histological examinations of the lesions have revealed inflammation and necrosis of the epidermis, and in severe areas the damage can penetrate as far as the upper dermis (Lynn *et al.,* 1991). The acute inflammation and necrosis typically seen in "burnt hocks" is probably caused by prolonged contact with corrosive substances in the litter (Bray, 1984), and as such, damage is likely to increase the longer birds are housed on poor litter. Strain differences in the reaction to corrosive substances are not thought likely, but there may be differences between strains in the rate at which corrosive substances are formed in the litter, and the amount of time birds spend sitting. As litter nitrogen has been implicated in the development of burnt hocks (Tucker and Walker, 1992), differences between strains in their requirements for nitrogen and essential amino acids for growth and maintenance will be of interest.

Whilst the dietary supply of essential amino acids and nitrogen is lower in Label Rouge type rations than in broiler rations, and nitrogen excretion is expected to be lower, the litter will be used for a longer period than in broiler production and the cumulative accretion of litter nitrogen may be important. This will be especially true if the litter surface becomes wet or greasy, as faeces accumulate on the upper surface.

Producers may be encouraged to feed alternative vegetable protein sources to soya for reasons of sustainability and product traceability, but the protein quality is expected to be poorer than for soya. Work at Gleadthorpe showed that birds were more likely to be downgraded due to hock burn when protein quality was poor (Tucker and Walker, 1992). This is because nitrogen excretion was increased and the litter was wetter as birds drank more water in order to excrete the nitrogen. In addition, antinutritive factors present in the plant material may either influence the availability of amino acids, (e.g. tannins), or metabolites may be toxic to the birds (e.g. metabolites of glucosinolates), and under these conditions litter moisture and nitrogen contents could be increased.

Accentuation of the breast musculature appears to protect the broiler from developing breast blisters, and to a lesser extent from broken clavicles (Moran 1999 citing Moran 1996). If birds are more active in extensive production systems then they may spend less time sitting, and if feather cover due to cold exposure is good then the detrimental effects of a reduced breast musculature may be reduced. Coarse and hard-packed litters should be avoided.

Bruising is the presence of blood in tissues and is caused by trauma. Bruises are unsightly and the meat is downgraded. They are mostly not detectable in the live bird and only become visible after feather removal at processing. Bruising occurs 12 to 24 hours before slaughter (Hamdy *et al.,* 1961) and most bruises occur on the breast, followed by the legs and wings, then the backs and thighs (Warris *et al.,* 1999). Well-muscled birds, which are heavy for their age, and females, have been claimed to be more susceptible. Whilst differences in the incidence of bruising between strains was reported by Taylor and Helbecka (1968), other authors have thought it unlikely (Griffiths and Nairn, 1984). However, if strains respond differently during handling, either at catching or shackling, and preliminary work at ADAS Gleadthorpe has identified more extreme behavioural responses in slower growing strains than in broiler hybrids, then the incidence of bruising may be affected.

Dietary regime has been implicated as a factor influencing carcass bruising (Moran and Stilborn, 1996). Broilers fed diets marginally deficient in crude protein had lower breast meat yields and increased carcass fatness than birds fed diets having adequate crude protein contents. Bruising tended to be greater in birds fed diets marginally deficient in crude protein.

Northcutt *et al.,* (2000) suggested a technique for identifying the age of a bruise based on the changes in colouration with time. If bruising is a greater problem in extensively produced table birds than in broilers, then it would be possible to apply their technique so as to identify when bruising occurred. This would enable improved *ante mortem* handling practices to be developed.

The accumulation of fat in the skin, and particularly in association with the main feather tracts, can act to minimise the effect or perception of traumas from handling prior to slaughter (Moran, 1999). A decrease of this fat layer along the pelvic back results in an accentuation of bruising. Conversely skin tears that occur during processing usually appear along feather tracts and a decrease in fatness reduces their likelihood.

Kafri *et al.,* (1985) found that feeding diets having a wide energy to crude protein ratio reduced skin breaking strengths in broilers, and skin breaking strengths were more affected at day 56 than at younger ages. The observed differences in skin breaking strength were not consistently associated with fat, protein, moisture or collagen concentrations of the skin. Later work by Kafri *et al.,* (1986) found that broilers fed diets having a wide energy to crude protein ratio had thicker and weaker skin. This was related to an increased thickness of the hypodermis, as more fat was stored in the adipocytes, and a less thick layer of dermis plus epidermis. The dermis is a compact, protein-rich fibrous layer composed largely of collagen but, also containing substantial quantities of elastin and glycoproteins (Kafri *et al.,* 1986 citing Spearman, 1971). Kafri *et al.,* (1986) did not consider the epidermis to be important in contributing to skin breaking strength because of its thinness. The authors suggested that the failure to relate skin breaking strength to its chemical composition in earlier work (Kafri *et al.,* 1985) may have been due to an inappropriate expression of their results or, that they were attempting to oversimplify the cause-effect relationships for skin integrity.

Several authors have found skin breaking strengths to be greater in male broilers than in female broilers (Edwards *et al.,* 1973; Kafri *et al.,* 1985; Smith *et al.,* 1977; *Kafri et al.,* 1986). Kafri *et al.,* (1986) related this to a reduced skin thickness, and most notably a thinner hypodermis in the males.

The findings suggest that there may be a greater likelihood of skin tears at slaughter in free range table birds than in intensive broilers as free range table birds are fed diets having a wide energy to protein ratio. Furthermore, when females only are used for free range table bird production the likelihood of skin tears at slaughter will be greater.

The susceptability of traditional free range table birds to skin tears at slaughter will depend on the appropriateness of the feed energy to crude protein ratio of the diet to the birds' growth profile. Slow growing birds will have a lower protein requirement for maintenance and growth than fast growing broiler hybrids. However, there is anecdotal evidence that skin tears during processing are greater in strains of slow growing birds than in broilers. This may be due in part to mechanical damage caused by the processing equipment having been developed and set for broiler-shaped birds, as opposed to slower growing strains which tend to have longer legs.

EFFECTS OF AMBIENT TEMPERATURE AND PHOTOPERIOD

Some environmental factors can influence litter moisture content when using a given dietary regime, and when litter moisture content exceeds 400 g/kg the likelihood of birds developing hock burn damage is greater (Tucker and Walker, 1992). Condensation occurs on a surface (and that can include litter) when the temperature of that surface falls below the dew-point temperature, which in turn is determined by the moisture content of the air. One of the functions of poultry house insulation is to keep the inside surface temperatures above dew-point, and for UK broiler houses a thermal conductance of better than 0.35 W/m^2 has often been recommended (Charles *et al.,* 2002). Housing facilities for extensively produced table birds may not be insulated to this standard, particularly very small scale houses, and even if they are, cold air entering through the popholes will lower the temperature within the house. The risk of condensation is greater when house temperatures are low and humidity is high, and in small scale housing facilities such conditions may be experienced during the night-time. In cold weather, producers are often tempted to reduce ventilation rates to conserve heat, but this profoundly affects air moisture content. Litter management is likely to be more difficult in small scale free range housing facilities as natural ventilation systems offer less control than automated ventilation systems, and options to provide heat may be limited and expensive. Litter is likely to require forking so as to break-up the compacted surface and new bedding may need to be added at frequent intervals.

Water spillage from bell drinkers or old fashioned bucket drinkers may also be greater in small scale housing facilities as 'water saving' nipple drinkers are unlikely to be suitable for use during cold weather. Particular attention will need to be paid to the litter-area underneath the drinkers as wet capped litter tends to spread out from the drinking source.

There has been relatively little work examining the effect of photoperiod on carcass condition, but work by Renden *et al.*, (1992) identified a higher incidence of breast blisters in birds receiving long daily dark periods. They suggested that this was because birds spent more time lying down, but reduced breast musculature may have been a contributing factor.

The incidence of haemorrhages, visible after slaughter, in broiler thigh muscles, is higher when grown in low temperatures (Kranen *et al.*, 1998). This is related to changes in haemodynamics and metabolic adaptations to an increased need for energy and oxygen at low temperatures. Broiler hybrids fed Label Rouge diets may be less susceptible to thigh muscle haemorrhages at low temperatures than broilers fed high specification diets. This is because the metabolic demands for growth will be less than in faster growing broilers.

There is a greater likelihood of litter being wetter and less friable in extensive table bird production than in intensive broiler production. This is because of weather penetration through popholes, and less water efficient drinkers may be used in extensive table bird production. Wet litter will increase the risk of hock burn damage and dirty birds in extensive table birds and, there is a risk of seasonal peaks in hock burn damage and dirties with the incidence possibly peaking during cold, wet and windy weather. Although the hocks of the birds may be removed at processing, hock burn affects the appearance of the dressed bird and hock removal is likely to be less acceptable for extensive table birds than for intensive broilers.

If catching is more difficult in extensive table birds than in intensive broilers then there is a risk of increased bruising and downgrading in extensive table birds. Downgrading due to bruising is very costly to the producer as birds have been grown to slaughter age and, in systems requiring long growing periods the financial loss is great.

Skin colour

Perhaps one of the most important quality attributes of a food product is its appearance. Appearance affects whether a food product is purchased or not, and colour is a major contributing factor (Fletcher, 1999). Once a product has been selected, colour is still important as it affects the perception of other sensory attributes.

The perceived colour of a product depends on three components. These are pigmentation, light source and the observer. Pigments absorb and reflect light of differing wavelengths, and so colour will be affected by the spectral distribution of the light source. The ability to observe colour will depend on the sensitivity of the observer to the reflected light wavelengths.

Skin colour of broilers is cream, or if the birds have been fed maize then it is yellow. The vast majority of broilers in UK are not maize fed and this reflects UK consumer demands. However, some consumers may think that a yellow skin colour is achieved by adding synthetic pigments to the feed and this may be off-putting. It should be made clear to consumers that yellow skin colouration in "corn-fed" chickens is achieved by natural means. This will particularly important for extensively produced table birds that have been "corn-fed", as consumers are unlikely to pay extra for premium poultry meat if it is thought that synthetic pigments have been fed to the birds.

It is difficult to know how UK consumers would react to skin colours other than cream or yellow, and yet there is a wide range of skin colours that are achievable through genetic and dietary means. The exposure to cream skinned broilers over very many years may make consumers reluctant to accept table birds having slate or black skin. However, it would be desirable for producers to be able to provide visible points of difference between niche table birds and broilers. This would enable a link to develop between the appearance of the table bird and specific eating qualities, such as a stronger flavour than intensive broiler meat. The chosen skin colour characteristic must be unique to table birds in order for it to form a basis for product differentiation. Yellow skin pigmentation is not specific to one production system, and so it would not be useful in this context. Factors affecting skin pigmentation and colour are discussed below.

Effects of breed and diet specification
Skin colour is dependent on the genetic ability of the bird to produce melanin pigments in the dermal, or epidermal melanophores, and the genetic ability of the bird to absorb and then deposit carotenoid pigments in the epidermis (Fletcher 1999). Black skin colour is produced when melanin pigments are deposited in both the dermis and epidermis, whereas blue skin colour (slate) is produced when melanin is deposited only in the dermis. A green skin colour may be produced when melanin is deposited in the dermis and xanthophylls are deposited in the epidermis. In the absence of melanin deposition, a yellow skin colour is produced when xanthophylls are deposited in the epidermis. Fletcher (1999) reported that in most commercial broiler strains the ability of the birds to produce melanin pigments have been eliminated through breeding programmes. With the exception of Cornish Game, traditional English breeds lack the ability to deposit carotenoid pigments in the skin, thus skin colour is white in appearance regardless of dietary xanthophyll content.

Details of skin colour in several medium to slow growing French commercial strains, as used in the study at ADAS Gleadthorpe are given in Table 1. A cream or yellow skin means that birds are able to deposit xanthophylls in the epidermis.

Table 1. SKIN COLOUR IN SEVERAL HUBBARD ISA STRAINS

Strain	Skin characteristics
Redbro	Cream/yellow
Master Gris	Cream/yellow
Gris Barre	Cream/yellow
I457	Cream/yellow
I657	White

There are two ways of achieving xanthophyll intake and deposition in extensively produced birds. The first is to provide birds with a diet rich in xanthophylls. The second, is to provide pasture species rich in xanthophylls, so that the consumption of pasture leads to pigmentation.

Formulating diets so as to achieve a desired level of pigmentation in extensively produced table birds will not be easy, as feed intakes will vary with changes in ambient temperature. In unexpected cold spells, feed intakes will increase and xanthophyll intakes may be excessive. Very intense yellow pigmentation is likely to be less acceptable to the consumer than paler coloured skins.

The pre-intensive egg industry used yellow maize and grass meal as dietary sources of pigment (Robinson, 1948), and occasionally lawn clippings (Thompson *et al.,* 1952).

Maize is the preferred cereal for feeding to domestic birds as its dietary energy value is the highest amongst cereals (Larbier and Leclercq, 1992). Maize is rich in xanthophylls, and the colouring ability of the xanthophylls is relatively high (Table 2). Grass meal has a low nutritive value and was abandoned mainly because it occupied too much dietary space. Its metabolisable energy (ME) value for poultry has been estimated at only 5 MJ/kg (McDonald *et al.,* 1995). Jonnson and McNab (1983) reported a linear reduction in the ratio of broiler live weight gain to feed intake as dietary grass meal content increased within the range of 0 to 125 g/kg. Skin colour became increasingly yellow as the dietary grassmeal content increased, and they suggested that when feeding diets containing 75 g/kg grass meal the skins would be too yellow for the UK consumer.

Table 2. XANTHOPHYLL PIGMENT CONTENTS OF RAW INGREDIENTS (ppm)

Raw material	Total xanthophylls (ppm)	Biological efficacity (colouring ability)	Major xanthophylls
Maize gluten	300	65	Lutein, zeaxanthine
Maize	25	80	Lutein, zeaxanthine
Lucerne meal	260	50	Lutein, zeaxanthine
Lucerne concentrate	1000	50	Lutein, zeaxanthine

Source: Larbier and Leclercq, 1992

Products derived from lucerne have been used for some considerable time in poultry diets, primarily as a source of xanthophylls (Table 17). The major disadvantage of lucerne is its low dietary energy value (for young birds the AMEn of lucerne meal is 4.6 MJ/kg DM, and similarly for lucerne concentrate 11.6 MJ/kg DM), but in addition lucerne meal contains saponins which have antinutritional activity (Larbier and Leclercq, 1992). Thus, the dietary energy value of lucerne meal decreases as the inclusion rate increases. Although lucerne concentrate has a very high crude protein content, and it is rich in lysine, tryptophan and threonine, a high level of saponins restricts its inclusion into compound diets to a maximum of 5%. However, where slower growth is required, as in extensive table bird production, it may be possible to use slightly higher inclusion rates. The availability of xanthophylls in lucerne is lower than in maize (Table 18), and variability may be induced by processing.

Grass may be a natural source of xanthophylls for extensively produced birds, and birds on range do appear to ingest grass. The feeding value of the pasture, if any, and its pigment value, must presumably depend upon the stage of growth, plant nutritional status and composition of the foliage, the grass species, and the degree of poaching and contamination. Many, or all of these may be affected by pasture and grazing management (see Chapter 7), so that questions arise concerning the utilisation of the pasture by the birds, the rotation and manurial practice used for the grass, and the species sown and surviving the grazing and trampling process.

McDonald *et al.,* (1995) cited work by Waite and Sastry (1949) which show a fall in the carotene content of the grass species Timothy from 274 mg/kg DM in May to about 66-88 mg/kg DM during July. Moisture contents were about 78% and 65%, respectively, and so on a fresh basis the carotene concentration of grass is low, compared with dried grass meal.

Grass consumption in birds fed wheat based rations is thought unlikely to be sufficient to produce an intense yellow skin. However, very little is known about the effects of voluntary grass intake on skin colour. It may be possible to develop sward compositions that are rich in xanthophylls, but grazing, poaching and seasonal effects on sward xanthophyll content are likely to affect its efficiency for pigmentation. Furthermore, variability in pasture intake between birds is likely to be high and this will reduce the uniformity of skin pigmentation.

It is exciting that there is the potential to produce table birds having different skin pigmentation characteristics to that of broilers. It enables breeder companies to expand what they have on offer to producers and retailers, and ultimately to the consumer, but perhaps more importantly it offers a potential means for product differentiation at the point of sale. Specific skin pigmentation characteristics, possibly 'dotted' pigmentation as found when removing coloured pin feathers, may be acceptable, or even appealing when buying more 'rustic' food products, such as traditional free range table birds.

It seems that yellow skin pigmentation may be more variable in extensively produced table birds than in broilers. This is because the diet may supply more xanthophylls than required for optimal skin colour when feed intakes are increased at lower than expected temperatures, or conversely, xanthophyll intakes may be less than optimal when feed intakes fall at higher than expected temperatures. Furthermore, even when producers are not aiming to produce table birds having yellow skin colour, there may be some differences in colour from cream through to yellow, because of differences in grass intake.

EFFECTS OF PHOTOPERIOD

Interestingly, pigmentation due to xanthophyll deposition in the epidermis is affected by the light environment (Fletcher *et al.*, 1977; Janky *et al.*, 1980; Janky *et al.*, 1985). Shank skin colour is brighter in birds grown in open-type houses than in birds grown in windowless houses (i.e. conventional broilers). It is possible that the enhanced skin pigmentation seen in birds exposed to sunlight, may confer some kind of protection against the sun's harmful rays.

Lastly, bird health is critical to uniform pigment absorption and deposition in the epidermis. Disease, particularly coccidiosis, has been shown to have dramatic negative effects on skin pigmentation (Fletcher, 1999).

Poultry meat colour

Poultry meat colour may differ from pale tan to pink in the raw meat and from tan to pale grey when cooked. Unpleasant meat colour, either in the raw or cooked product, may lead to consumer rejection. The major factors contributing to meat colour are myoglobin content, the chemical state and reactions of myoglobin, and muscle pH (Fletcher, 1999). Myoglobin content has been primarily related to species, muscle (e.g. thigh meat, versus breast meat) and age of bird. Muscle type and age of bird are relevant in this review and they are discussed below. The chemical state and reactions of myoglobin with other compounds greatly influence meat colour. Muscle pH following *rigor mortis* onset affects both the light reflectance properties of the meat and the chemical reactions of myoglobin. Higher muscle pH is associated with darker meat, whereas lower muscle pH is associated with lighter meat (Fletcher, 1999a; Fletcher, 1999b; Fletcher *et al.,* 2000; Wilkins *et al.,* 2000). Raw breast meat colour and pH affects cooked breast meat colour and pH but, cooking reduces the degree of colour variation (Fletcher *et al.,* 2000).

There is accumulating evidence that darker coloured breast meat has a shorter shelf-life than lighter coloured breast meat, and that this may be due to a higher pH in darker breast meat (Allen, 1997). Objectionable odours occur sooner in dark fillets than in light fillets, and dark fillets have higher numbers of *Psuedomonas* spp. *Psuedomonas* spp. are one of the primary bacteria producing off-odours in spoiled poultry meat (Allen, 1997 citing Pooni and Mead, 1984). *Shewanella putrefaciens* is the other primary spoilage species isolated from broiler carcasses, and although growth is inhibited at normal breast meat pH they proliferate at higher pHs (Allen, 1997, citing Newton and Gill, 1981). Spoilage of poultry meat is due to psychrotrophic bacteria degrading muscle amino acids (Pooni and Mead, 1984), and it should not be confused with rancidity, where unsaturated fatty acids are peroxidised.

The relationships between meat colour, pH and shelf-life are highly relevant to extensively produced poultry meat. Retailers will want to maximise the shelf-life of a premium product, particularly if the product is purchased in 'leisure mode', whereby sales patterns may be less consistent than for staple food products. Furthermore, spoilage will be much more expensive than for a staple food product. If strain differences in meat pH occur (Scheurs *et al.,* 1995, see later discussion on cooked poultry meat texture and tenderness), then spoilage rates for traditional poultry meat may differ from those of broiler meat.

Although most extensively produced birds are currently sold whole with skin-on, it is thought that there will be a future expansion in the portioned market. Variation in breast meat colour as seen between skinless broiler fillets sold in multipacks is likely, but when buying a premium product uniformity may be more important. Wilkins *et al.,* (2000) reported large variations in broiler breast meat colour between flocks, but minimal variation within flocks. The reasons for flock variations in breast meat colour are not known.

MUSCLE MYOGLOBIN CONTENTS, AND EFFECTS OF AGE AND STRAIN

Breast meat is paler in colour than thigh meat and this is reflected in the widespread use of light and dark meat as descriptive terminology for muscle origin. The paler colour of breast meat is related to its lower myoglobin content than that of dark meat. Miller (1994) reported breast meat myoglobin contents of only 0.01 mg/g meat, and dark meat myoglobin contents of 0.40 mg/g meat, when sampled at eight weeks of age. Other workers have been unable to quantify breast muscle myoglobin content when broilers were killed and sampled at only six weeks of age (Kranen *et al.,* 1999).

Muscle myoglobin contents increase with age (Miller, 1994), and this is associated with darker, or more redder meat. Miller (1994) reported increases in chicken breast muscle myglobin content from 0.01 mg/g at 8 weeks of age, to 0.10 mg/g at 26 weeks of age. Other workers have reported an increase in darkness and redness of turkey meat with age of bird (Froning and Hartung, 1967).

Strain differences in turkey meat colour were reported by Froning and Hartung (1967), and it is suggested that this may have been due to strain differences in the rate of maturation. Kranen *et al.,* (1999) reported that muscle myoglobin concentrations correlate with muscle fibre type composition. If this is true, then differences in muscle fibre composition at a given age may influence meat colour.

Extensively produced table birds are older than broilers at slaughter and so, muscle myoglobin concentrations will be higher and, the meat will be redder than in broilers. If redder poultry meat does have a higher pH than whiter poultry meat, then extensively produced table birds may have a shorter shelf-life than broilers.

The *post mortem* conversion of muscle to meat and a discussion on muscle pH decline is given later.

RED OR PINK COLOUR PROBLEMS IN COOKED POULTRY MEAT

Pinkness in fully cooked poultry meat is a major defect. Ahn and Maurer (1990) reported that haem-complex forming reactions of myoglobin and haemoglobin produced orange-red to pink colouration of strong intensity. Histidine, methionine, cystine, or their side chains from the solubilised proteins, and vitamin B_6 derivatives are important factors in the pinkness of cooked turkey meat. A recent review of the primary factors influencing pink discolouration in cooked white meat is that by Maga (1994).

Nutrient composition

Demby and Cunningham (1980) found large differences in published nutritional values for chicken meat, and this was largely accounted for by differences between studies in breed, feed, age, method of production, sex, processing and type of meat sampled. The protein content of raw broiler meat ranged from 170 g/kg (citing work by Cook, 1974, using broiler pieces) to 233 g/kg in white meat (citing work by Millares and Fellers, 1948). The fat content of white meat ranged from 21 g/kg (citing McCance and Widdowson, 1960) to 102 g/kg (citing Goodwin and Simpson, 1973), and in the dark meat from 10 g/kg (citing Wladyka and Dawson, 1968b) to 141 g/kg (citing Goodwin and Simpson, 1973).

Holland *et al.,* (1991), in the current edition of McCance and Widdowson's classic tables of compositions of foods, on which work began in 1926, and which were first published in a form resembling the present publication in 1940, provide analyses for 1188 foods, including 14 chicken items. All of the items were analysed to establish the concentrations of water, protein, fat, carbohydrate, nitrogen, saturated, monounsaturated and polyunsaturated fatty acids, cholesterol, starch, sugars, 12 minerals and 14 vitamins. The energy values of the products were also measured. Supplements to McCance and Widdowson's tables include those by Paul *et al.,* (1980) on amino acid and fatty acid compositions, and Chan *et al.,* (1995) with detail on poultry meat. Extracts from these sources give some of the data of particular interest in relating poultry products to human requirements (Tables 3 to 5).

Poultry meat is a useful source of high quality protein, and of many vitamins and minerals, and it is a particularly rich source of riboflavin, niacin, vitamin B_6, phosphorus, manganese, zinc and selenium (Charles, 1996). Provided that chicken is eaten without the skin it is low in fat (Department of Health, 1994).

EFFECTS OF AGE, SEX AND BREED

Goodwin and Simpson (1973) found higher moisture and lower fat contents in very young broilers marketed as Rock Cornish hens, than in older broilers, but protein contents were similar. Other workers have reported differences in breast muscle B vitamin contents with age (Singh and Essary, 1971). Riboflavin content was higher at four weeks of age than at six, eight or ten weeks of age, and niacin and thiamine contents were higher at six weeks of age than at four, eight or ten weeks of age.

Male broiler carcasses had higher moisture and protein contents, and lower fat contents than female broilers (Goodwin and Simpson 1973). Riboflavin contents were higher in meat from males than in meat from females, but both sexes had approximately the same concentrations of niacin and thiamine (Singh and Essary, 1971).

Edwards and Denman (1975) reported that body compositions varied greatly among breeds of chicken, with certain breeds having consistently higher nutritional values than others.

The findings suggest that the meat from free range table birds and traditional free range table birds may be higher in fat than broiler meat, and that niacin and thiamine contents may be lower than in broiler meat. There are likely to be interactive effects of age and sex of bird on the nutrient composition of chicken meat and this will be relevant to free range table birds when females only are used. Females have higher meat fat contents than males

Table 3. SOME MAJOR NUTRIENT CONTENTS IN POULTRY MEAT (g/100g) AND THE ENERGY VALUE OF POULTRY MEAT (kJ)

Sample	Water	Protein	Fat	Energy (kJ)	Fatty acids			Cholesterol
					Saturated	Monounsaturated	Polyunsaturated	
Total chicken meat	75.1	22.3	2.1	457	0.6	1.0	0.4	90
Breast meat	74.2	24.0	1.1	449	0.3	0.5	0.2	70

Table 4. SOME MINERAL AND VITAMIN CONTENTS OF POULTRY MEAT

	Calcium (mg/100g)	Iron (mg/100g)	Zinc (mg/100g)	Selenium (mg)	Vitamin B_6 (mg/100g)	Folate (mg)	Vitamin D (mg)	Thiamin (mg/100g)
Total chicken meat	6	0.7	1.2	13	0.38	11	0.10	0.14
Breast meat	5	0.5	0.7	12	0.51	14	0.2	0.14

Table 5. ESSENTIAL AMINO ACID CONTENTS IN CHICKEN MEAT (mg/100g)

	Isoleucine	Leucine	Lysine	Methionine	Cystine	Phenlyalanine	Tyrosine	Threonine	Tryptophan	Valine	Histidine
Total chicken meat	950	1540	1840	490	260	920	720	850	230	980	620

Sources: Holland *et al.*, (1991); Paul *et al.*, (1980); Chan *et al.*, (1995)

whereas meat riboflavin contents are lower in females than in males. The extents to which these effects are exacerbated by age are not known. It is thought that differences in nutrient compositions between breeds are likely to be in fat content and, contents of those minerals and vitamins that are stored in fat. Breeds that are slower growing and less efficient at converting nutrients into lean meat will be fatter. It would be interesting to see if there are breed differences in meat nutrient contents when birds are slaughtered at similar metabolic ages.

EFFECTS OF DIET SPECIFICATION

Diet specification can influence the nutritional value of poultry meat. The total fat content of the carcass, and fat quality are both issues of concern to consumers. Factors affecting abdominal fat deposition, this being the major fat depot in poultry, have already been discussed, and they included feed energy to protein ratios, and post brooding environmental temperature. However, consumers may not understand that fat accumulates in the abdomen, and its presence in a dressed table bird may raise suspicions.

A report on public attitudes to food safety highlighted consumer concerns about what chickens were fed, and about the methods of modern poultry-raising (Food Standards Agency, May 2000). Many respondents thought that there were implications for human health associated with poultry production. Whilst most people believed that free range chickens were healthier to eat and tasted better than intensively produced birds, there was growing scepticism about what "free-range" meant and many respondents strongly suspected that reality did not accord with what the label suggested. From this report, it does not seem that labelling and certification schemes will instil confidence in consumers. Thus, if a niche product fails to meet the consumers' perception of what it should be, then the product will not be purchased again.

Consumers have sometimes complained that broiler carcass fat is too runny during cooking, and this reflects its high polyunsaturated to saturated fatty acid ratio. It is possible to alter the ratio of polyunsaturated to saturated fatty acids in the carcass by altering the ratio in the feed. However, since a high polyunsaturated to saturated fatty acid ratio is considered nutritionally desirable, it would not seem sensible to deliberately saturate the carcass fat. Furthermore, the use of saturated fats (e.g. tallow) in poultry rations may not be acceptable to the consumer, and in organic systems of poultry meat production it is not permitted.

The essential fatty acid profile of poultry fat reflects the essential fatty acid profile of the feed. Linoleic acid (c18:2 *n*-6) and α-linolenic acid (c18:3 *n*-3) are essential fatty acids for poultry (Enser, 1999). The bird can convert them into longer, more unsaturated fatty acids, such as eicosapentaenoic (EPA 20:5 *n*-3) and docosahexaenoic acid (DHA 22:6 *n*-3), respectively (*loc. sit.*), but the conversion of α-linolenic to EPA may be limited (Olomu and Baracos, 1991).

The deposition of α-linolenic acid in poultry tissues is advantageous as it can lower the 18:2 to 18:3 (n-6:n-3) ratio in the human diet which should increase the synthesis of EPA and DHA in humans. A ratio of *n*-6 to *n*-3 of five, or less is recommended for the human diet (Department of Health, 1994).

Yan *et al.,* (1991) found that differences in specific fatty ratios in the breast tissue reflected differences in these ratios in the dietary oils, but adipose tissue more closely reflected the lipid composition of the diet than did muscle tissue. Hargis and van Elswyk (1993) cited the findings of Smith (1991), who suggested that this was because lipids stored in adipose tissue have a non-functional role, whereas lipids serve a functional role in muscle tissue. Cell membranes comprise phospholipids, and DHA is present at higher concentrations in phospholipids than in neutral lipids, whereas α-linolenic acid is present at relatively low concentrations in phospholipids (Enser, 1999 citing Lin *et al.,* 1989).

There are some indications that omega-3 fatty acids are preferentially deposited in muscles, and in particular breast meat (Hulan *et al.,* 1988). However, as dark meat contains about twice as much lipid as white meat, both types of meat are expected to provide similar quantities of EPA and DHA per 100 g serving (Enser, 1999).

Soya bean meal is the predominant protein source for poultry, and if there are restrictions on permitted feed ingredients, such as in organic production where solvent extracted products are not allowed, then there will be a very high dependence on full-fat soya. Other feed ingredients such as expelled rapeseed meal, or blends of extruded rapeseed and pulses may be used in small quantities during the finisher stage, but a low confidence in the birds' ability to tolerate antinutritive factors during early life generally precludes them from starter rations. Soya is rich in linoleic acid and fairly rich in oleic acid, and sunflower is similar to soya, whereas, linseed is a rich source of α linolenic acid (Table 6). Rape is rich in oleic acid and fairly rich in linoleic acid.

Table 6. FATTY ACID COMPOSITION OF OIL SEEDS (% OF ETHER EXTRACT)

	Oleic C18:1	*Linoleic C18:2*	*Linolenic C18:3*	*Others*
Soya	22.6	50.2	7.1	20.1
Rape	57.7	21.6	9.0	11.7
Sunflower	20.3	64.9	0.2	14.6
Linseed	18.5	16.1	51.6	13.8

Source: van Kempen and Jansman (1994)

It appears that there is reasonable scope to alter both the ratio of linoleic to α-linolenic acid in the diet and in poultry tissues, and that this may be achieved by reducing the proportion of soya within the diet. If vegetable diets are required then a suitable replacement may be a blend of linseed and pulses, possibly with some rape. However, the ability to alter the amount of DHA and EPA in poultry tissues when feed ingredients are restricted to vegetable sources seems to be more limited.

There are implications for flavour and rancidity of poultry meat associated with manipulating the dietary fatty acid composition. Flavour will be discussed later in the text.

Sterrif *et al.,* (1977) reported that dietary vitamin E supplementation increased liver and lipid α-tocopherol concentrations in poultry and this retarded the development of carcass rancidity. The amount of vitamin E required in the diet depended on the type and amount of fat fed. Other workers have also found increased carcass stability when feeding higher levels of vitamin E (Bartov and Bornstein, 1981).

Rancidity occurs when fats and oils are oxidised and the processes involved in oxidation were recently described by Adams (1999). Fat oxidation is initiated and propagated by various free radicals, whereby the oxidisable substrate is converted into an alkyl radical by proton, or electron abstraction.

The final stage, known as termination, comprises the recombination of various species of free radicals to produce stable end products (e.g. hydrocarbons, aldehydes, ketones, alcohols and organic acids), and it is the end products that generate characteristic unpalatable, or rancid flavours and odours, which make the food unacceptable to the consumer.

Antioxidants function by modifying the oxidative reactions in one of three ways: chain breaking, oxygen scavenging, or interference with the initiation of oxidation (Adams, 1999). Tocopherol is a "chain-breaker"; it intercepts free radicals involved in the auto-oxidation process. Free radicals of antioxidant molecules are formed in this process, but the new free radicals are less reactive and they do not react with lipid to form new free radicals, and so the oxidative chain reaction is broken.

The γ and δ forms of tocopherol are more effective antioxidants in fats and oils than α tocopherol, but α tocopherol appears to be the most important form in human nutrition (Adams, 1999). The regeneration of vitamin E following chain-breaking, is thought to occur through a reaction with vitamin C. Both vitamin E and vitamin C need to be in the diet of humans as they are not manufactured by body tissues. Vitamin E also reacts with selenium, and both are important in maintaining the body's defence mechanisms (e.g. antibody production, cell proliferation, cytokine production, prostaglandin metabolism, and the non-specific immune system).

Poultry meat may provide a valuable source of tocopherol (Table 7), and there is the potential to increase tissue concentrations of tocopherol by diet manipulation (Enser, 1999 citing the work of King *et al.*, 1995).

Table 7. TOCOPHEROL CONCENTRATION OF SOME FOODS (mg/100g)

Food	Tocopherol (mg/100 g)
Broccoli	0.6
Cabbage	7.0
Carrots	0.6
Chicken	4.0
Potatoes	<0.1
Rapeseed oil	57 - 89
Sunflower oil	49 - 80
Tomatoes	0.4

Source: Adams (1999)

The high energy to protein ratio of rations fed to extensive table birds mean that table birds are likely to be fatter than intensive broilers. There is some evidence that the fatty acid composition of chicken fat may be altered by dietary means and so there is scope to make the fat healthier. This will be more readily achieved in adipose tissues than in intramuscular fat within the breast muscles. However, reducing the ratio of linoleic to α-linolenic within the fat of table birds may reduce their shelf-life as the fat will be more susceptible to auto-oxidation. The dietary supply of anti-oxidants to table birds will need to be optimised so as to delay rancidity and maintain shelf-life.

References

Note that the references for Chapters 4, 5 and 5 are listed at the end of Chapter 6.

6

CONVERSION OF MUSCLE TO MEAT, FLAVOUR AND TENDERNESS. DISCUSSION OF MEAT QUALITY ISSUES

Conversion of muscle to meat

Lyon and Buhr (1999) reviewed the biochemical basis of meat texture in poultry. The *post mortem* depletion of energy reserves within the muscle initiates the onset of rigor and the demarcation between muscle and meat. The degree of muscle contraction at the onset of rigor is variable, and it can be influenced by several physiological pathways and processing procedures. It is the main determinant of meat tenderness (Lyon and Buhr, 1999 citing Locker, 1960).

The myofibrillar proteins actin and myosin comprise over half of the protein of skeletal muscle and they are major factors in the *post mortem* onset and the resolution of rigor, tenderness, water holding capacity, emulsification and binding properties of meat (Lyon and Buhr, 1999). The cessation of blood flow at the time of death results in the depletion of intracellular oxygen and initiates the consumption of all cellular energy reserves (ATP and creatine phosphate). *Post mortem* regeneration of ATP requires cellular metabolism to change from aerobic to anaerobic glycolysis and this causes the rapid utilisation of muscle glycogen and the accumulation of lactic acid as a waste product. Muscle pH falls from about 7.2 (Dransfield and Sosnicki, 1999) to a metabolic ultimate of pH 5.4, below which glycolysis is inhibited (Lyon and Buhr, 1999). The magnitude of the pH decline and the quantity of lactic acid accumulated *post mortem* depends on the amount of muscle glycogen present at the time of death. Preslaughter handling (such as prolonged food withdrawal, or physiological stress during transport) can deplete muscle glycogen stores and affect the rate at which they are broken down after death, thus influencing the rate and extent of acidification (Warris *et al.,* 1999). It seems that leg muscle glycogen stores are affected most.

Post mortem ATP keeps muscles extensible and there is little change in muscle length. This is known as the 'delay phase' prior to the onset of rigor

(Warris *et al.,* 1999). The authors cited work by Khan (1975) who identified proportional ATP usage as being the trigger for the onset of rigor. In the subsequent rapid phase of rigor there is abrupt sarcomere shortening and a reduction in muscle extensibility. The pH at the onset of the rapid phase is linearly related to the ultimate pH (Lyon and Buhr, 1999 citing Bate-Smith and Bendall, 1949). Minimal changes in muscle tension occur within the pH range 6.7 to 6.4, but muscle shortening and the development of tension occurs rapidly as the pH drops below 6.3, and peak tension occurs at pH 6.0 (Lyon and Buhr, 1999 citing Newbold, 1966). The last phase of rigor is resolution, or postrigor, when a muscle is again extensible, but in contrast to the onset of rigor, muscles in this phase remain extended after loading (Lyon and Buhr, 1999). Resolution of rigor is due to the degradation of muscle ultrastructure.

The onset of rigor in broilers is about two to four hours *post mortem* (Lyon and Buhr, 1999). In poultry, the ultimate pH for predominantly white or predominantly red fibre type muscles differs. Red leg muscles reach an ultimate pH within 2-3 h *post mortem* (pH 6.0-5.9), whereas white breast muscle pH may continue to decline beyond 24 h (pH 5.6-5.4) when carcasses are rapidly chilled *post mortem*. The longer periods for rigor onset and pH decline in white fibre type muscles are due to greater initial energy reserves, mainly in the form of glycogen (Lyon and Buhr, 1999 citing Hay *et al.,* 1973). *Rigor mortis* also resolves earlier in leg muscles than in breast muscles (Lyon and Buhr, 1999 citing Kijowski *et al.,* 1982).

Not all fibres within a muscle shorten uniformly *post mortem* (Lyon and Buhr, 1999). One reason for this is variation in muscle composition (fibre type). Furthermore, even when a muscle is composed entirely of a single fibre type (e.g. *M. pectoralis major*) the degree of shortening throughout the muscle at any time *post mortem* may not be uniform due to differences in fibre width at specific locations within the muscle (Smith and Fletcher, 1988), or due to variable energy reserves and their utilisation (Lyon and Buhr, 1999). Thus, toughness of meat may not correlate directly with mean sarcomere shortening, but with an increased incidence of highly shortened sarcomeres which form a rigid continuum.

Tenderness can be affected by both very slow and very fast rates of early *post mortem* glycolysis (Lyon and Buhr, 1999). The rate of glycolysis is affected by muscle temperature. A low temperature retards the rates of glycolysis and lactic acid production and this reduces the rate of pH decline, whereas a high temperature accelerates pH decline (Lyon and Buhr, 1999 citing Marsh, 1954). In the latter case, there is greater fibre shortening and increased toughness. By lowering the carcass temperature fibre shortening is resisted and this results in increased tenderness of meat aged-on-the-bone. However, this is not the case when carcasses are exposed to cold temperatures whilst ATP is still present in the muscle cell; that is prior to *rigor mortis*, as sarcomeres shorten in a process termed "cold shortening" (Sams, 1999 citing Hamm, 1982). Poultry breast muscle is less prone to cold shortening than red muscles, but not entirely exempt (Biligi *et al.,* 1989).

Removing table birds from houses at slaughter age may be more difficult than for broilers. Catching ease will depend on physical factors such as site location, size and layout of the houses, lighting facilities and flock size. It will be more difficult to remove birds that are reared in small scale facilities set some distance from proper access roads. Articulated lorries having modular draw units are unlikely to be able to access remote sites, and it may be necessary to crate birds and move them by tractor and trailer from the site to the lorry. Extensively produced table birds may be more active than broilers, and this may making catching more difficult. Stressful catching conditions will deplete muscle glycogen stores *ante mortem* and this will adversely affect the rate of pH decline and meat tenderness *post mortem*. Table bird producers using mobile houses must ensure that birds can be easily removed and crated if meat quality is to be optimised.

Cooked poultry meat texture and tenderness

In a review, Lyon and Buhr (1999) described various processing procedures having direct effects on the texture of cooked poultry meat. This chapter will consider the effects of production factors on cooked meat texture.

EFFECTS OF AGE, BREED AND DIET SPECIFICATION

Eating quality traits differ with age and breed. Poultry meat tends to be less tender as birds age (e.g. Shrimpton and Miller, 1960) and in red meat this has been related to an increase in fibre diameter (Herring *et al.,* 1965). As cross sectional growth occurs to a greater extent in fast-glycolytic fibres (type 2b) than in other fibre types (Moran, 1999, citing Remignon *et al.,*

1995), and the *M. pectoralis major* is composed entirely of fast-glycolytic fibres (Smith and Fletcher 1988), there is the potential for increased toughness in premium breast meat.

Lyon *et al.,* (1984) examined the texture profile of hot stripped breast meat (that is meat removed pre rigor, followed by rapid freezing and then slow thawing to produce a thaw rigor) from male and female broilers at 7, 9, 10 and 11 weeks of age. Age affected all textural attributes (springiness, chewiness, hardness and cohesiveness). Hardness at 11 weeks of age was greater than at 7, 9 or 10 weeks of age. Cohesiveness declined as age increased, whereas chewiness tended to increase with age. Males had higher chewiness values than females. Springiness increased with age and males had higher springiness values at 7, 9 and 11 weeks of age, but at 10 weeks of age there were no sex differences.

As extensive table birds are older at slaughter the meat will be firmer, more springy and more chewy, but less cohesive than broiler meat. There will be differences in meat texture between male table birds and female table birds, the meat from males being more chewy. It is likely that sex differences in meat texture will be greater for traditional free range birds than for free range birds, as traditional free range birds are older at slaughter. This means that the more expensive product, the traditional free range table bird, will be have more variable meat textures than the cheaper free range table bird, and than the much cheaper broiler.

As discussed previously, muscle fibre size and muscle maturation rates differ between meat-type breeds and egg-type breeds (Goldspink and Yang, 1999), and so breed choice will affect meat texture. Breeds having slower growth profiles than modern broiler hybrids will be required for use in traditional free range and organic production systems, as the minimum age at slaughter is 81 days, and it would be possible to delay the attainment of market live weight in broiler hybrids only by applying quantitative feed restriction techniques. Even if this were to be acceptable, work by Shrimpton and Miller (1960), which used a factorial design for assessing the effects of feed restriction and strain-type on toughness in poultry meat, found denser meat textures in feed restricted fast growing birds (White Rocks) than in full fed slow growing birds (Red Leghorns). Feed restriction reduced meat tenderness because when birds were full fed the meat texture of fast growing birds was preferred.

The interactive effects of genotype (Ross 1 broiler, *versus* ISA 657), diet and stocking density, as used in either conventional broiler production or Label Rouge production, on the sensory attributes of breast and thigh meat were examined by Farmer *et al.,* (1997). After processing, sample birds were placed in roasting bags and they were cooked in computer controlled ovens (oven temperature 190°C) until the internal temperature of the breast muscle reached 90°C, as measured by a thermocouple probe. Panellists scored breast and thigh meat texture on a linescale 100 mm long; a high score indicating a high intensity for the attribute being measured. Overall acceptability provided a measure of the tasters' preferences.

There were interactive effects of strain and diet on toughness and resistance to cutting by knife. Ross 1 fed a commercial broiler ration had tougher breast meat than Ross 1 fed a Label Rouge ration, and tougher breast meat than ISA 657 fed either a commercial broiler ration, or a Label Rouge type ration.

The findings were similar for resistance to cutting by knife. It is reasonable to assume that the higher concentrations of crude protein and essential amino acids in the broiler ration would have not only promoted faster whole body growth in Ross 1, but also promoted increased muscle fibre diameter, compared with birds fed a Label Rouge diet.

If poultry muscle fibre diameter is related to toughness, as in bovine studies (Herring *et al.,* 1965), then the findings of Farmer *et al.,* (1997) are not surprising. By comparison, the Label Rouge ration may have been better suited to the growth requirements of ISA 657, and the extra protein provided in the commercial broiler ration may not have been used for increased muscle growth. The authors reported a greater degree of fibrousness in ISA 657 than in Ross 1. This may be due to more dense packing of smaller fibres in ISA 657. There were no effects of strain, or diet on the remaining textural attributes, and there were no effects of stocking density on any of the textural attributes assessed.

Other reasons why breast muscle tenderness may differ between strains have been considered. Schreurs *et al.,* (1995) reported lower *post mortem* ultimate pHs in breast muscles of strains selected for fast growth than in breast muscles of an egg-type strain (slower growing White Leghorn). A low ultimate pH and a rapid rate of pH decline are associated with increased sarcomere shortening and an increase in the susceptibility of myosin to denaturation. Thus, the likelihood of producing tough, pale breast meat with a poor water holding capacity may be greater in fast growing strains (Dransfield and Sosnicki, 1999).

Post mortem proteolysis is thought to weaken the muscle fibres, and this may lead to tenderisation. Dransfield and Sosnicki (1999) reported that muscle concentrations of cathespins (lysosomal enzymes) and calpains (calcium dependent proteolytic enzymes) were higher in slow growing strains than in fast growing strains. This may better enable tenderisation in extensively produced poultry meat.

EFFECTS OF AMBIENT TEMPERATURE AND PHOTOPERIOD

The authors are not aware of any work examining the effects of ambient temperature, or photoperiod on meat texture. However, there may be effects of ambient temperature on muscle fibre size and muscle lipid content, and these may alter meat texture characteristics (firmness, fibrousness, stringiness and juiciness). For example, at high ambient temperatures feed intake may be reduced, live weight gain may be reduced and breast meat yields may be lower, whereas carcass fatness may be increased (Yahav *et al.,* 1996). Breast muscle fibre diameter may be reduced and the meat texture may be firmer and more fibrous than in birds grown under thermoneutral conditions.

The effects of photoperiod on meat texture are perhaps less easy to postulate as there may be unknown seasonal influences, additional to photoperiodic effects on growth rate and muscle fibre growth.

EFFECTS OF EXERCISE

The potential for exercise to modify muscle growth, dimension and fibre type proportions has already been discussed, but there is also some evidence suggesting that meat texture may be affected by exercise. If pasture design is to be developed so as to optimise pasture usage, then the effects of exercise on meat texture will be of increasing interest.

According to Skaarup (1983, cited by Farmer *et al.,* 1997) free range birds have less tender breast meat than birds reared in confinement, but thigh meat is little affected. Sandusky and Heath (1988b) found effects of pen design on breast muscle growth, and their findings may offer some insight into why breast meat texture may be affected by exercise. A single-sided ramp placed within a pen produced birds with faster growing breast muscles than birds in conventional pens. Breast muscles after four weeks of exposure to the single-sided ramp were longer. It is suggested that if exercise increases breast muscle length, due to increased muscle extension, then muscle fibre dimensions and meat texture may be affected.

Extensively produced table birds exit the house through popholes and down ramps onto the pasture. Ramp usage may produce differences in breast meat texture between extensively produced table birds and broilers. Presumably the steepness of the ramp will affect the amount of work needed to walk up the ramp and the extent to which the breast muscles are extended. Variations in ramp steepness may affect breast muscle growth, muscle fibre dimensions and meat texture. Also there may be differences between birds in the number of times they go outside and this may lead to variation in meat texture within a flock.

Ducks kept on pasture have been reported to have more red muscle fibres, and the diameter of red and white muscle fibres were smaller than those kept intensively (Dransfield and Sosnicki, 1999 citing Pingel and Knust, 1993). Dransfield and Sosnicki (1999) suggested that smaller diameter fibres may allow more dense packing and increased toughness of the meat.

The effects of ranging activity on breast and leg muscle dimensions, muscle fibre widths and lengths, and meat texture are poorly understood, and work should be aimed at providing a better understanding.

TEXTURAL CHANGES DURING COOKING AND COOKING METHODOLOGY

Textural changes in meat during cooking are closely related to the crosslink characteristic of its constituent collagen (McCormick, 1999). Morphologically there are three discrete collagen depots in muscle, the epi-, peri-, and endomysium. The connective tissue sheath surrounding individual muscles, and continuous with the tendon joining other muscles or bones, is the epimysium. The epimysium is often thick and tough, and resistant to both shear and solubilsation. However, it is easily and usually separated from cuts of meat and is generally not considered to be a factor in meat quality. The three-dimensional collagen network that surrounds large and small bundles of muscle fibres and contains intramuscular lipid deposits and vasculture is the perimysium. The layer of connective tissue encircling each muscle fibre and overlying the basement membrane is the endomysium. Intramuscular connective tissue is the combined perimysium and endomsyium depots. The vast bulk of the intramuscular connective tissue, about 90%, is the perimysium. The perimysium is thought to play the major role in determining meat texture differences that are related to connective tissue (McCormick, 1999 citing Light *et al.,* 1985). The role of the endomysium in meat texture is less well understood.

McCormick (1999) summarised the work of Davey and Gilbert (1974), where temperature dependent increases in shear force (toughening) were reported and a basis for the role of collagen in determining cooked meat texture was given. Using beef *Sternomandibularis* muscle, a biphasic increase in the toughness of meat occurs as temperature increases. The first sharp increase in toughness occurs between 40 and 50 °C, and this corresponds with the denaturation of the myofibrilar proteins actin and myosin. Before a temperature of 60°C is reached, both these proteins undergo a transition from the gel state, initially posing little resistance to shear, to a hardened dehydrated form, although water is still associated with the myofibre. At 64 to 68°C, there is a second sharp increase in toughness, and this corresponds with the denaturing of collagen. Thermal denaturation of collagen associated with fibrils results in shrinkage of the fibril and a commensurate increase in force, or tension. The degree of tension or shrinkage developed is a function of how heavily the collagen is crosslinked with mature, heat stabile crosslinks (McCormick, 1999 citing Bailey and Light, 1989). As temperature increases from 80 to 90°C collagen is eventually gelatanised and toughness diminishes. Other authors have associated changes in meat toughness during cooking with connective tissue orientation, actin-myosin overlap and water content (Lyon and Buhr, 1999 citing Currie and Wolfe, 1980).

In skeletal muscles, reducible collagen crosslinks are rapidly replaced with mature non reducible forms (McCormick, 1999). The author pointed out that the progression of crosslinking occurred significantly faster in avian skeletal muscle than in mammals. Perhaps this reflects differences in the rate of development between species. McCormick (1999) cited the work of Velleman *et al.,* (1996) who reported that the highest concentrations to date of non reducible collagen crosslinks were found in breast muscles taken from White Leghorns at one year of age.

In general, locomotor muscles possess more crosslinked collagen than postural muscles, and exercise influences crosslinking patterns. However, the mechanisms responsible for regulating crosslink formation are not known. Multiple factors affect muscle growth and adaptation over time, and McCormick (1999) suggested that when these factors result in connective tissue turnover, and usually muscle growth, collagen characteristics will also be markedly influenced.

As extensively produced table birds are older at slaughter, the amount of crosslinked collagen in the meat may be greater than in broilers, and at cooked temperatures of less than 80°C this may make the table bird meat tough. Extensively produced table birds are also likely to have been more active than broilers and this will affect the pattern of collagen crosslinks in the leg meat. It is suggested that the effects of cooking method and final cooking temperature on the texture of table bird meat should be examined.

Chicken meat flavour

Flavour is one of the main eating quality attributes which, together with appearance and texture, dictates our choice and enjoyment of foods (Farmer, 1999). Flavour is a combination of the sensations perceived by the two chemical senses, taste and smell. Taste is perceived by the taste buds on the tongue and other parts of the mouth, and they mainly detect the four principal tastes: sweet, sour/acid, salt and bitter. Other tastes that can be detected are astringency, metallic, pain ('hot' and 'cooling' foods) and 'deliciousness'. The sense of smell detects certain chemicals which stimulate the olfactory receptors at the top of the nasal cavity. Odours are detected in the air above the food before it is eaten and they may also be detected during eating as they pass in the breath from the mouth, through the posterior nares at the back of the nose, into the nasal cavity.

In a review on flavour formation, Farmer (1999) listed compounds that are thought to contribute to the taste, or aroma of cooked chicken meat. Amino acids, peptides, proteins and nucleotides are most important for the taste of red meat and the author suggested that similar results might be expected for poultry. Compounds which contribute to taste in chicken are generally present in the raw meat and do not require cooking for their generation. However, studies on red meat have shown that cooking may affect the concentrations of taste substances, and similar effects would be expected for chicken meat. In addition, changes in sugars, amino acids and nucleotides which occur during cooking will affect aroma and flavour, since many of these substances are also precursors for the chemical reactions responsible for the formation of odour compounds.

About 500 volatile aroma compounds have been reported for cooked chicken, but many of these compounds have high odour threshold levels and make little contribution to the overall flavour (Farmer, 1999). Farmer listed 34 key odour compounds. The compounds include a range of sulphur compounds, heterocyclic compounds containing oxygen or nitrogen, and aldehydes and keytones. Individually these compounds confer 'sulphurous', 'meaty', 'toasted', 'roasted', 'fatty', 'tallowy', 'fruity', or 'mushroom' aromas, but together they combine to give the characteristic aroma of cooked chicken. In contrast to red meat, lipid oxidation products are important to the aroma of cooked chicken (Farmer, 1999 citing Gasser and Grosch, 1990), and seven of the key odour compounds are aldehydes and ketones produced from n-6 and n-9 fatty acids (Farmer, 1999 citing Grosch 1982). A further two key odour compounds, these being lactones, are produced by the oxidation of triglycerides.

Although triacylglycerols are not needed for the development of poultry meat flavour they can affect juiciness (Keeton, 1993) and the release of flavour (Cheevance and Farmer, 1997, cited by Farmer, 1999), and therefore they may influence how flavour is perceived.

The Maillard reaction (reaction of amino acids, or peptides and proteins with reducing sugars) and the degradation of thiamine (vitamin B_1) are also important in the development of aroma compounds. Of the 34 key odour impact compounds listed by Farmer (1999), 20 of them may be formed by the Maillard reaction. The precursors of the Maillard reaction present in raw meat are derived from the degradative reactions which occur in muscle after death. Free sugars and some sugar phosphates are formed by glycolysis and by the breakdown of ATP (Lawrie, 1992).

There are interactions between the flavour and aroma of cooked chicken and cooking method, but little attention has been given to establishing differences in key odour impact compounds between cooking methods (Farmer, 1999). This may become increasingly important, especially if there is an expansion into portioned premium poultry meat, as this increases the range of cooking options available to the consumer.

EFFECTS OF AGE, BREED AND DIET SPECIFICATION

Poultry meat flavour tends to be enhanced after the growth inflexion point on the Gompertz curve (Moran, 1999, citing Touraille *et al.,* 1981ab, and Scholtyseek and Sailor, 1986). Knizetova *et al.,* (1991) evaluated the Gompertz function for nine broiler lines and observed that the age of inflexion varied between 48 and 56 days for males and between 48 and 53 days for females. Larbier and Leclercq (1992) published inflexion points of 48 days and 43 days for male and female broilers, respectively, these being several days earlier than those published by Knizetova *et al.,* (1991). Typical slaughter ages for broiler hybrids grown in an intensive production system are 39 to 44 days, whereas the minimum permitted age at slaughter in a free range system is 56 days. Thus, flavour may be stronger when broiler hybrids are grown in a free range system provided that the growth inflexion point is not excessively delayed by dietary crude protein dilution. Work at ADAS Gleadthorpe showed that the age at growth inflexion of as hatched Ross 308 broilers was delayed from about day 39 to about day 55 by feeding Label Rouge diets, as opposed to feeding higher specification diets (Table 7). As growth inflexion occurs earlier in females than males, and because of the close proximity of the age at growth inflexion to the minimum slaughter age for free range birds, the flavour of meat from females may be more pronounced than for males.

Many commercial producers use female broiler hybrids in a free range system as they have a growth profile better suited to the longer growing period. Furthermore, the males may be used in intensive broiler production where their faster growth rates and better feed conversion efficiencies provide financial benefits. If consumers expect free range birds to have a more intense flavour than broilers, then the use of female broiler hybrids in free range systems seems to be most appropriate. An example of work examining chicken meat flavour at ages relevant to conventional broiler production and free range production is that of Sonaiya *et al.,* (1990). Breast and thigh meat flavours were more intense in birds slaughtered at day 54 (similar to free range production), than in birds slaughtered at day 34.

Larbier and Leclercq (1992) published a growth inflexion point of about 53 days in male Label Rouge traditional free range birds. This was similar to the age at growth inflexion for ISA 657 fed high specification diets, but much earlier than for ISA 657 fed Label Rouge type diets (day 50 and days 65/66, respectively, Gleadthorpe data see Table 7). However, in either case, growth inflexion occurred many days prior to the minimum age at slaughter (day 81), and so the flavour of traditional free range table birds is expected to be enhanced.

Farmer (1999) cited a series of experiments by Touraille *et al.,* (1981a, b) in which the flavour intensity of male Label Rouge chickens was found to increase up to 14 weeks of age. This was attributed to physiological changes that were occurring as the birds approached sexual maturity. At this stage, there are differences in muscle growth and development between the sexes, notably thigh muscle development in cockerels (Iwamato *et al.,* 1993a citing Ono *et al.,* 1982, 1983), and increased lipid deposition in females (Iwamato *et al.,* 1993a citing Iwamato and Takahara, 1971 and Iwamato *et al.,* 1975). Ricard and Touraille (1988, cited by Farmer, 1999) reported that flavour intensity was similar between males and females until they reached 14 weeks of age, at which time the males had more intense flavoured breast and thigh meat. Farmer (1999) suggested that the effects of sex on flavour are related to age, and the stronger flavour achieved when males reach maturity.

As traditional free range birds may be approaching sexual maturity at the minimum slaughter age of 81 days, there is a risk that flavour intensity may differ between sexes. Whilst consumers are likely to purchase traditional free range chickens because they want to experience eating more flavoursome meat, they may not expect variations in flavour intensity. Either an increase, or a decrease in flavour intensity may be viewed unfavourably by the consumer. Increased flavour intensity may surprise the consumer, and depending on their preferences, they may prefer a more 'gamey' flavour, or they may dislike the stronger flavour. A decrease in flavour intensity on repeat buying may lead the consumer to doubt whether birds really have been grown according to the purported production system, especially if there is a perception that access to pasture and fresh air adds flavour. A recent study on consumer attitudes to food purchasing, funded by the Food Standards Agency, illustrated that consumers were doubtful of whether free range birds were allowed to range (May 2000).

Aside from the economic advantage of slaughtering at the minimum slaughter age, because feed conversion efficiencies are falling as the birds get older, it may also be wise for producers to avoid delaying slaughter in order to help minimise flavour variations. Although, in theory it would be possible to market males and females separately, so that the males are sold as more intensely flavoured, this would be yet another labelling complication. However, it may be possible for a subjective flavour scoring system to be set up, such that the intensity of flavour is described on a point scale used by the consumer to select the appropriate product. Subjective flavour scores have been developed and applied to describe the intensity of cheeses, coffees and wines at point of sale.

Flavour enhancement is thought to be due to physiological changes that occur as the chicken grows older, and these changes affect the presence of taste compounds and flavour precursors (Farmer, 1999). It would be useful to examine birds prior to and post growth inflexion when attempting to elucidate chemical reasons for age-related flavour enhancement.

Farmer (1999) noted that whilst there have been many studies examining the effects of genotype, or strain, on chicken meat flavour, the findings have been inconsistent. However, in studies where birds have been tasted at the same weight, strain effects have mostly disappeared. In other studies, flavour differences between strains have been associated with increased age at market weight. It is suggested that a relevant basis for examining flavour differences between strains would be to make the comparisons at a similar metabolic age. Strains having a faster growth rate are likely to have a higher mature live weight than slower growing strains. If comparisons are made at similar live weights, then the metabolic age of the slower growing strain will be higher than that of the faster growing strain.

Diet manipulation may offer the greatest potential to enhance poultry meat flavour. Triglycerides and short chain fatty acids are oxidised during cooking to form either aldehydes, ketones or lactones and these products contribute to the aroma of cooked poultry meat (Farmer, 1999 citing the work of Gasser and Grosch, 1990). Polyunsaturated fatty acids (PUFAs) are more reactive than saturated fatty acids in oxidative reactions, and so manipulating the PUFA composition and content in poultry fat is likely to offer a greater potential for generating desirable meat flavours.

Enser (1999) reported that although the effect of different dietary fatty acids in poultry has not been related to positive effects on flavour, in cattle and sheep, grass feeding produces stronger flavoured meat, as preferred by UK consumers, whereas feeding cereals produces a milder flavoured meat. The difference in flavour is generally attributed to whether or not the main dietary fatty acid is ∝-linolenic (18:3 *n*-3), which produces strong flavour, or linoleic acid (18:2 *n*-6), which produces a mild flavour (Melton, 1990). Enser (1999) suggested that as poultry normally have high levels of linoleic acid in their tissues this may partly contribute to their mild flavour. Cereal and soya based rations, as commonly used by the UK poultry industry, have relatively high linoleic acid contents, and especially when soybean oil, or maize oil are used as the main fat source. When attempting to lower the dietary ratio of linoleic acid to ∝-linolenic acid, rapeseed oil is marginally better than soybean oil as its linoleic acid content is lower. However, the ∝-linolenic acid content of linseed oil is much greater than in that of soybean oil or rapeseed oil (e.g. typically 500 g/kg compared with less than 120 g/kg), and this makes it a potentially useful ingredient. Alternatively, some of the soybean meal component of the diet may be replaced by vegetable protein sources that have a lower linoleic acid content, for example peas, or field beans. (Extruded full fat soya has a linoleic acid content of 10.2 g/kg, whereas peas and field beans have linoleic acid contents of 0.52 g/kg and 0.45 g/kg, respectively; Larbier and Leclercq, 1992). Home grown protein sources may be traceable and this confers advantages over imported soya. In addition, there would be sustainability advantages associated with the use of home grown protein sources in poultry rations.

Rancidity and off-flavours in PUFA-enriched poultry meat may be overcome by supplementing the birds' diet with antioxidants, and various forms of tocopherol are likely to play an important role (e.g. Ajuyah *et al.,* 1993). Several herbs and spices (rosemary, sage, oregano, thyme, allspice, cloves, curcumin, ginger, mace, nutmeg and turmeric) have a powerful antioxidant capacity (Adams, 1999). However, the effect of feeding herbs, or spices at a biologically active concentration on meat flavour and shelf life need to be examined.

EFFECTS OF AMBIENT TEMPERATURE AND PHOTOPERIOD

Temperature may affect flavour through changes in muscle lipid content and lipid composition, but only a small number of studies have reported temperature effects (reviewed by Farmer, 1999). We are not aware of any published work examining the effects of photoperiod, light intensity, or spectral composition on meat flavour, but it is likely that any effects would be through changes in feed intake, growth rate and maturity, and fatness. Advancing sexual maturity through seasonal increases in photoperiod may influence meat flavour in males grown in a traditional free range production system.

EFFECTS OF PASTURE USAGE

Flavour may also be modified by access to range and the consumption of herbage, or insects, on range. Mead *et al.,* (1983) reported that birds consuming a higher proportion of whole wheat and fresh green vegetables had increased intestinal counts of *E. coli* and faecal streptococci. Their breast meat flavour differed from that of the control birds and was described as being richer, meatier, sweeter or more roasty, but sometimes gamey, or "off". If herbage or insect intake on range modifies the gut microflora then there is the potential for meat flavour to be affected. It is noted that the composition of vegetative pasture used in extensive poultry meat systems has been designed to meet the producers' requirements for ease of management and durability, short resilient grasses being the major component. Herbs, or other vegetation may be worthwhile pasture components if they provide positive flavour enhancements when consumed at modest levels.

Although grass consumption contributes to increased meat flavour in ruminants through differences in fatty acid composition (Enser, 1999) and thermal oxidation products, grass intake in poultry is not thought likely to be sufficient to alter the fatty acid composition of poultry lipids. This is because the fat content of fresh grass is very low and consumption would need to be very high if it were to impact on the composition of total fatty acids consumed.

EFFECTS OF COOKING METHOD

Farmer (1999) reported on unpublished work that examined the effect of cooking method on meat flavour. Open roasting gave higher roast aromas and roast flavours than when birds were cooked either in a roaster bag, roasting dish or foil-wrapped. The lowest sensory scores were given to foil-wrapped birds, whereas scores were similar for birds cooked in a roaster bag, or a roasting dish.

It would be appropriate to examine the effects of cooking technique and final internal cooking temperature on poultry meat aroma and flavour. Whole traditional chickens are likely to be purchased by consumers having an interest in both cooking and food, and part of the pleasure of cooking will be anticipating eating the meal and experiencing the aromas whilst it is cooking.

Discussion

Many attributes contribute to the overall quality of a table bird and they have been discussed above. Both the number of quality attributes, and the complex inter-relationships between growth and many eating quality traits, makes optimisation of table bird quality a challenge. Furthermore, when a consumer purchases a whole bird for roasting, or pot-roasting, many muscles contribute to the total amount of edible meat, and all of the meat is expected to perform well under one cooking method. This is quite different to when consumers purchase premium beef, pork, or lamb products as they are sold according to specific body parts.

It will be difficult to embark on work aimed at optimising table bird quality without a sound understanding of consumer preferences and requirements. Although it is possible to list many quality attributes that will be required regardless of eating quality preferences, such as dressed weight, freedom from skin blemishes, tears, bruises, hock burn, breast blisters and odours, these quality attributes are common to both broilers and table birds, and so they will not influence extra expenditure for a premium product. Meat colour, flavour and textural qualities will be important in influencing the consumers' purchasing decision.

As there has been a long period over which UK consumers have been exposed to broiler meat, a system of subjectively scoring poultry meat texture and flavour intensity at point of sale may be useful. A subjective description of texture and flavour would enable consumers to choose new and alternative products that fit their own eating preferences. In addition, the application of a subjective scoring system for texture and flavour, and the linking of this to information gained by retailers offering customer loyalty cards, would enable eating quality preferences to be determined at a very basic level. However, a more useful approach would be to use focus groups to establish consumer preferences. It would seem logical to bias the focus groups so that members are chosen on the basis that they are likely to purchase premium poultry products (income groups A and B; life stages 3, 5 and 6).

It is thought that the positive promotion of extensive table bird production within the UK will be fundamental to the success of premium poultry products. The Label Rouge accreditation system, set up in France in the 1960s to change the public perception of some foods, is now widely accepted by French consumers as offering superior products to conventional. In 1992, 77.8 million birds, representing one third of the chickens traded as whole carcasses, were produced under the 'Label' accreditation scheme (Culioli *et al.,* 1994). The success of this scheme is exemplary, and in particular the French consumers' belief in the product. By contrast, many UK consumers appear to be sceptical of farming practices within the UK, and the labelling of products as free range may not be sufficient to win consumer acceptance. The public's image of free range production may not tally with reality, and common criticisms are that birds do not range. Aesthetics of production will be important, and the development of pastures and management techniques that lure the birds onto the range are likely to be promotionally desirable.

If product labelling or accreditation systems are not believed by the UK consumer then product authentication based on analytical techniques may be an option. Fumiere *et al.,* (2000) recently reported that near infrared (NIRS) reflectance spectroscopy may be used to discriminate between slow growing chickens and broilers with an accuracy of between 80 and 100%. The authors suggested that the technique may be suitable for integration into an analytical system of surveillance of certified poultry meat products. Visible differences between table birds and broilers, such as 'rustic' skin pigmentation in table birds may also help to encourage a belief that the product is different to broilers.

It seems that there are certain quality characteristics that will be determined by the rules of a given production system, and in particular, specifications for minimum age at slaughter, and minimum dietary cereal contents. The minimum age at slaughter, and the need to attain a family-sized roast at this age, will dictate how suitable a breed is to a production system. However, growth rate is not an isolated variable, as there are many quality traits that are influenced by growth rate. These include important eating quality traits, such as meat texture and flavour, and possibly meat colour. Poultry meat texture is linked with muscle fibre size, and at slaughter age this will relate to the metabolic age of the bird as affected by nutrition and environment. At slaughter age, fast growing strains are likely to have larger muscle fibre diameters than slower growing strains, and this may increase resistance to cutting, or chewing. Conversely, slower growing birds may have more densely packed narrower muscle fibres, and this may make the meat feel stringey or fibrous. In either case, the meat texture will differ from that of broilers.

Flavour will undoubtedly be more intense in older birds. This is due to the deposition of flavour precursors in the muscle as the birds get older, and several workers have linked this to around the time of growth inflexion on the Gompertz curve. An understanding of the processes behind age-related flavour development may enable strategies to be developed that aim to optimise the taste and aroma of cooked poultry meat.

The widespread use female broiler hybrids in free range production, although based on the need to achieve dressed birds within a specific weight band, seems sensible as females have an earlier growth inflexion, and flavour may be stronger than in males. Although the continued selection for faster growth rates in broiler hybrids may not be antagonistic to flavour development, the birds' growth profile may become less suited to free range production. It should be remembered that broiler hybrids are developed for the broiler industry and that the supply of chicks to free range producers is a relatively small area of business for the broiler companies.

Broiler hybrids are not suited for use in traditional free range production as their growth rate is too fast, and feed restriction would be needed. This would not be in keeping with an image that the niche market must promote. Traditional UK breeds of poultry may offer some potential in the future, perhaps for consumers wanting very intensely flavoured birds, or for consumers wanting local speciality food products. However, for the present, it is thought that slow growing hybrids will be most suited for use in traditional free range production in UK.

There is some evidence in the reviewed literature of breed differences in the activity of muscle proteolytic enzymes, and in slower growing birds tenderisation may be better enabled. Furthermore, the decline in pH during the conversion of muscle to meat may be less than for broilers, and this may reduce sarcomere shortening and meat toughness. However, a negative aspect of this is that premium poultry meat may have a shorter shelf life than broiler meat, as a higher meat pH favours the growth of meat spoiling bacteria. Meat colour may be redder than in broiler meat, and if the portioning market develops for premium poultry meat, this should be taken into consideration at the point of sale.

Diet specification will affect the ratio of breast to thigh meat, and the deposition of lipids. Although consumers may accept a less plump bird, well-developed broiler breast muscles provide protection against the development of breast blisters and bruises, and in extensively produced birds carcass damage may be a problem. Flightiness and greater activity in free range birds may lead to a higher incidence of collisions. It is observed that when aircraft or large birds fly over the paddock the birds often run *en masse* back to the house. Pophole design and ease of access to the house will be important in preventing bruises, and yet to our knowledge there has been no research carried out on pophole design. Facilities within the paddock should also be optimised to provide shelter from aerial predators, as this would provide the birds with an alternative escape route. On-going work at Gleadthorpe is indicating that chickens like to sit under dense natural canopies, and these are being provided in the form of conifer pyramids.

There are interactive effects of breed, diet specification and temperature on feed intake, growth (whole body growth, allometric growth and muscle fibre size), fatness and eating qualities such as meat texture, and possibly flavour. However, the majority of work has been carried out using broiler hybrids grown to much younger ages.

Diet specification has less effect on whole body growth and breast meat yield at about day 56 in slower growing breeds, than in broiler hybrids, and this is not unexpected. Broiler hybrids have a fast early growth rate and their nitrogen and essential amino acid requirements for growth are high prior to growth inflexion (for whole body and breast muscle). Wheat is low in crude protein and essential amino acids, and so high dietary cereal contents, as stipulated for use in extensive table bird production, dilute the protein component of the ration. As the birds get older, the differential in live weights between birds fed high specification diets and birds fed Label Rouge rations is increasing by similar for different breeds.

Although the largest reduction in breast meat yield due to low lysine intake is for broiler hybrids, it may be undesirable to increase the dietary lysine content throughout the growing period as the birds may be too heavy at day 56. However, it would be interesting to examine the effects of increasing dietary lysine supply between day old and day 14, this being the time of rapid breast muscle growth, on live weight gain and breast meat yields.

Breast meat yields at about day 81 in Label Rouge birds seem to be little affected by diet specification. This gives some indication that the lysine supply of Label Rouge diets is not too deficient for breast muscle growth. However, there may be a case for examining the effects of dietary methionine and cystine supply on feather cover and feed conversion efficiency. Dietary methionine contents in extensive table bird rations are about half that of broiler rations, and yet feather cover will be more important in free range birds in maintaining body temperature than in broilers housed in controlled environments.

Feed intake and the birds' energy and nutrient requirements for maintenance will be affected by ambient temperature, and if they go outside in cold windy weather there will be a wind chill effect. Rations should be formulated so as to take into account the effects of temperature on feed intake, and this may help to reduce seasonal variations in live weight, meat yield and carcass fatness. However, it may be less easy to accommodate changes in the birds' energy and nutrient requirements during sudden unexpected hot or cold spells. Larger scale producers may be better able to respond, particularly if they mix their own rations, but smaller scale producers may only be able to respond only by adding, or reducing the quantity of whole cereals fed. Carcass fatness may vary with temperature conditions, and this may affect the 'finish' of the dressed bird, susceptibility to skin tears, meat juiciness and the perception of flavour.

It is difficult to assess seasonal effects on meat quality and eating quality. This is because, in conventional broiler production, seasonal effects have been largely ignored as birds are grown in windowless, controlled environment houses. However, birds grown in a free range production system have daytime access to pasture for at least 28 days, and daytime access is even longer in birds grown in a traditional free range system. At certain times of the year there are relatively large increases or decreases in natural daylength over a 28, or 42 day period, and despite the use of artificial lighting, chickens may respond to the natural daylength. If increasing natural daylengths advance sexual maturity in birds grown to day 81, then there may be differences in meat conformation, fatness and flavour intensity between males and females.

There is the potential to alter chicken meat flavour by altering the fatty acid composition of adipose lipids through dietary means. Unsaturated fatty acids are most important as they are oxidised during cooking and the volatile products contribute to the aroma of cooked meat. Work carried out in ruminants suggests that the ratio of linoleic acid to linolenic acid in the feed, and thence in the adipose tissue, has an effect on meat flavour. A high bias towards linoleic acid produces less flavoursome meat than when the ratio of linoleic acid to linolenic acid is reduced. In poultry rations, soya is the common source of protein, and even in Label Rouge starter rations it may be included at concentrations of up to 260 kg/t. Soya is rich in linoleic acid and less rich in linolenic acid, and so the ratio is highly biased towards linoleic acid. This bias may produce less flavoursome poultry meat than if alternative protein sources to soya were used, such as linseed, rapeseed, peas, or beans. The ability to modify meat flavour through dietary means is exciting and this will be examined further.

Utilisation of pasture by the birds may affect muscle size, muscle fibre type and meat texture. This may be through changes in muscle tension during exercise. The effects of exercise on musculature and meat quality in poultry have not been studied to any extent, and forced exercise as used in the limited number of studies available (the need to cross a barrier in order to gain access to feed) is likely to be qualitatively and quantitatively different from voluntary exercise in free range table birds. Access to pasture may also affect meat flavour, as herbage, or insect intake on range may modify the gut microflora. There is a real need to evaluate pasture compositions for poultry, and to study intakes, as pasture may be a valuable source of nutrients. Pasture compositions have been developed to be hard wearing, to withstand poaching and grazing, and to require relatively infrequent cutting. It is likely that, with prolonged use, such pastures would offer birds very little nourishment. A different approach may be to develop pasture compositions that are palatable to the birds, and rich in nutricines (as defined by Adams, 1999).

A move away from broiler production, although small at present, has re-iterated the importance of growth theory. Growth concepts were important in broiler production during the 1980s when ascites mortality and sudden deaths were high. Rapid early growth of the 'demand' organs (e.g. muscle) was thought to be a contributing factor, and so techniques aimed at slowing early growth were applied. This allowed the 'supply organs' (e.g. heart, lungs, liver) to develop, whilst delaying the growth of the 'demand organs'. In extensively produced table birds, growth concepts will be important as growth (whole body, allometric and muscle fibre growth, and metabolic age) will have effects on many quality attributes, including eating qualities.

In extensive production systems many factors are likely to vary: breed choice may differ between producers; the birds' environment will vary throughout the growing period; environmental conditions may differ across the UK; ranging activity may vary between birds; and, intakes of feed and pasture may vary between birds. This suggests that the qualities of premium poultry meat products may be more variable than those of broiler meat products. A better understanding of the complex interactive effects of breed, sex, diet, temperature and light on growth, meat quality and eating quality may enable some of the variation to be reduced.

It is an exciting time for the poultry meat industry and for consumers, as the number of potential new premium poultry meat products is set to expand.

Conclusions

BREED CHOICE

Breeds should be chosen so that their growth profiles accord with achieving suitable live weights at market age when fed diets meeting the statutory requirements for the production system, and when access to feed is *ad libitum.* Female broiler hybrids have growth profiles that produce birds of the required weight at day 56. They are not considered to be suitable for use in production systems requiring an 81 day growing period. Slow growing hybrids, as used in France in Label Rouge production, are thought to be suited to traditional free range and organic table bird production. Although none of the extensive poultry meat production systems require a growing period of greater than 81 days, it is thought that there will be a role for the traditional breeds in the niche product market.

GROWTH CONCEPTS

Growth concepts will be important in understanding the complex interactive effects of breed, sex, diet, temperature and light on poultry meat qualities and eating qualities. Most of the work reviewed in these chapters examined breed differences in meat qualities and eating qualities at live weight, or age reference points; but a more useful approach may be to compare qualities at similar metabolic ages.

CARCASS CONFORMATION

Carcass conformation will differ from that of broilers, and whilst this may be an acceptable feature of extensively produced table birds, there may be higher incidences of breast blisters and bruising. In addition, if there is an expansion into premium poultry meat portions, the amount of breast meat produced will be important.

NUTRIENT COMPOSITION

The nutrient composition of premium poultry meat, in particular its fat content, may be more variable than that of broiler meat. The fatty acid composition of the adipose tissue will reflect the fatty acid composition of the diet and, if there is a move away from soya based diets, the ratio of linoleic acid to linolenic acid in the adipose tissue may be reduced. This may increase the strength of poultry meat flavour, as unsaturated fatty acids are important in flavour development during cooking.

SKIN COLOUR

There are a wide variety of skin colours in poultry. Skin pigmentation characteristics may be developed such that they are recognised by consumers and associated with birds grown in extensive production systems.

MEAT COLOUR

Meat from slower growing birds may be redder than broiler meat, and if this is related in part to meat pH, then the shelf-life of traditional table birds may be shorter than that of broilers.

MEAT FLAVOUR

Flavour will be more intense in traditional table birds than in intensive broilers, and possibly more intense in free range table birds than in intensive broilers. The flavour intensity of males grown to day 81 may be greater than for females, as flavour intensity develops in the males at the time of sexual maturity. Seasonal increases in natural daylight may affect the development of sexual maturity, and thence flavour intensity.

MEAT TEXTURE

Meat texture will be firmer in traditional free range birds than in broilers, and this reflects breed differences in growth rate of muscle fibres. Factors affecting growth rate and muscle fibre size are likely to affect meat texture.

RANGING

Ranging may affect muscle characteristics, and in breast muscles the fibre length may be increased, whereas in leg muscles the fibre type proportions may be affected. This may influence meat texture, notably the perception of meat fibres, or tenderness; but the relative contribution of exercise to poultry meat eating quality is not known. Furthermore, some birds may range more than others, and some breeds may be better rangers. There may also be effects of wind, rain and sunshine on ranging. As ranging is likely to be very variable both within and between flocks, meat eating quality may vary.

PASTURE HERBAGE INTAKE

Pasture herbage intake and the contribution of pasture to the birds' total nutrient intake is likely to be variable, but low. This is because pasture compositions have been developed on the basis of easing pasture management, and being hard wearing. In extensive production systems, pastures should be regarded as potentially valuable sources of nutrients, and sward compositions should pertain to this ideal. Furthermore, herbage intake may intensify poultry meat flavour, although this will be dependent on intake being sufficient to modify gut microflora.

COOKING METHOD

The effects of cooking method and of final internal cooking temperature on the development of aroma in poultry meat has not received a great deal of attention. This may be an increasingly important component of eating quality work, as consumers buying premium products will want "the whole experience", and cooking and kitchen aromas will be components.

References
(for Chapters 4, 5 and 6)

Acar, N., Moran, E.T. Jr. and Mulvaney, D.R. (1993). Breast muscle development of commercial broilers from hatching to twelve weeks of age. *Poultry Science* **72**: 317-325

Adams, C. A. (1999). *Nutricines*. Nottingham University Press, Nottingham, UK

Ahn, D.U. and Maurer, A.J. (1990). Poultry meat colour: kinds of heme pigments and concentrations of the ligands. *Poultry Science* **69**: 157-165

Ajang, O.A., Prijono, S. and Smith, W.K. (1993). Effect of dietary protein content on growth and body composition of fast and slow feathering broiler chickens. *British Poultry Science* **34**: 73-91

Ajuyah, A.O., Hardin, R.T. and Sim, J.S. (1993). Effect of dietary full fat flaxseed with and without antioxidant on the fatty acid composition of major lipid classes of chicken meats. *Poultry Science* **72**: 125-136

Allen, C.D. Russell, S.M., and Fletcher, D.L. (1997). The relationship between broiler breast meat colour and pH shelf-life and odor development. *Poultry Science* **76**: 1042-1046

Arberle, E.D., Addis, P.B. and Shoffner, R.N. (1979). Fibre types in skeletal muscles of broiler- and layer-type chickens. *Poultry Science* **58**: 1210-1212

Ariele, A., Meltzer, A. and Berman, A. (1980). The thermoneutral temperature zone and seasonal acclimatisation in the hen. *British Poultry Science* **21**: 471-478

Bailey, A.J. and Light, N.D. (1989). *Connective tissue in meat and meat products*. Elsevier Science Publishers Ltd, Essex, UK

Bartov, I., Bornstein, S. and Lipstein, B. (1974). Effect of calorie to protein ratio on the degree of fatness in broilers fed practical diets. *British Poultry Science* **15**: 107-117

Bartov, I. and Bornstein, S. (1981) Stability of abdominal fat and meat of broilers: combined effect of dietary vitamin E and synthetic antioxidants. *Poultry Science* **60**: 1840-1845

Bartov, I. and Plavnik, I. (1998). Moderate excess of dietary protein increases breast meat yield of broiler chickens. *Poultry Science* **77**: 680-688

Bate-Smith, E.C. and Bendall, J.R. (1949). Factors determining the time course of rigor mortis. *Journal of Physiology* **110**: 47-65

Biligi, S.F., Egbert, W.R. and Huffman, D.L. (1989). Research note: Effect of postmortem aging temperature on sarcomere length and tenderness of broiler *Pectoralis major*. *Poultry Science* **68**: 1588-1591

Blunn, C.T. and Gregory, P.W. (1935). The embryological basis of size inheritance in the chicken. *Journal of Experimental Zoology* **70**: 397-414

Brackenbury, J.H. and Williamson, A.D.B. (1989). Treadmill exercise training increases the oxidative capacity of chicken iliotibialis muscle. *Poultry Science* **68**: 577-581

Bray, T.S. (1984). The effect of the diet on the litter condition and downgrading of broilers. ADAS internal report, Gleadthorpe EHF, Meden Vale, Mansfield, Notts, NG20 9PF, UK

Broadbent, L.A., Wilson, B.J. and Fisher, C. (1981). The composition of the broiler chicken at 56 days of age: output, components and chemical composition. *British Poultry Science* **22**: 385-390

Brody, S. (1945). *Bioenergetics and growth*. Reinehold, New York, USA

Cahaner, A., Pinchavasov, Y., Nir, I. and Nitsan, Z. (1995). Effects of dietary protein under high ambient temperatures on body weight, breast meat yield and abdominal fat deposition of broiler stocks differing in growth rate and fatness. *Poultry Science* **74**: 968-975

Chan, W., Brown, J., Lee, S.M. and Buss, D.H. (1995) Meat, poultry and game.5th supplement to the 5th edition *McCance and Widdowson's The composition of foods*. Royal Society of Chemistry and Ministry of Agriculture, Fisheries and Food

Charles, D.R. (2000). Egg purchasers - who are they and what do they want? BOCM PAULS Free Range Conference, Wakefield, 13 June

Charles, D.R., Clark, J.A. and the late Tucker, S.A. (2002) Ventilation rate requirements and principles of air movement. In: *Poultry environment problems: a guide to solutions*. Edited by Charles, D.R. and Walker, A.W., Nottingham University Press, Nottingham, UK

Charles, D.R., Groom, C.M. and Bray, T.S. (1981). The effects of temperature on broilers: interactions between temperature and feeding regime. *British Poultry Science* **22**: 475-481

Cheevance, F.F.V. and Farmer, L.J. (1997). Influence of fat on the flavour of an emulsified meat product. In: *Proceedings of Charalambous Memorial Symposium*, American Chemical Society, Washington, DC, Limnos, Greece, pp. 255-270

COMMISSION REGULATION (EEC) No 1538/91 of 5 June 1991 introducing detailed rules for implementing Regulation (EEC) No 1906/90 on certain marketing standards for poultry

Cook, D.J. (1974). The nutritional value of frozen foods. II. The composition of frozen foods. *British Nutrition Foundation* **12**: 42

Cowan P.J. and Michie, W. (1978). Environmental temperature and broiler performance: the use of diets containing increased amounts of protein. *British Poultry Science* **19**: 601-605

Crompton, D.W.T. and Walters, D.E. (1979). A study of growth of the alimentary tract of the young cockerel. *British Poultry Science* **20**: 149-158

Culioli, J., Touraille, C. and Ricard, F. (1994). Meat quality of "Label Fermier" chicken in relation to production factors. In: *Proceedings of the 9th European Poultry Conference*, Glasgow, 7-12 August, **2**: 25-28

Currie, R.W. and Wolfe, F.H. (1980). Rigor related changes in mechanical properties (tensile and adhesive) and extracellular space in beef muscle. *Meat Science* **4**: 123-143

Davy, C.L. and Gilbert, K.V. (1974). Temperature dependent cooking toughness in beef. *Journal of Science of Food and Agriculture* **25**: 931-938

Dawson, L.E., Walters, S. and Davidson, J.A. (1958). Cooked meat yields from four strains of chickens, 6 and 16 weeks of age. *Poultry Science* **37**: 227-230

Demby, J.H. and Cunningham, F.E. (1980). Factors affecting composition of chicken meat. *World's Poultry Science Journal* **36**: 25-67

Department of Health (1994). An action plan from the Nutrition Task Force to achieve the health of the nation targets on diet and nutrition. NTF1. DSS, Heywood

Dransfield, E. and Sosnicki, A.A. (1999). Relationship between muscle growth and poultry meat quality. *Poultry Science* **78**: 743-746

Edwards, H.M. and Denman, F. (1975). Carcass composition studies 2. Influences of breed, sex and diet on gross composition of the carcass and fatty acid composition of the adipose tissue. *Poultry Science* **54**: 1230

Edwards, H.M., Denman, F., Abou-Ashour, A. and Nugara, D. (1973). Carcass composition studies. 1. Influence of age, sex and type of dietary fat supplementation on total carcass and fatty acid composition. *Poultry Science* **64**: 2143-2149

Emmans and Fisher, C. (1986). Problems in nutritional theory. In: *Nutrient requirements of poultry and nutritional research*. Edited by Fisher, C. and Boorman, K.N., Butterworths, London, UK, pp. 9-39

Emmans, G.C. (1987). Growth, body composition and feed intake. *World's Poultry Science Journal* **43**: 208-227

Enser, M. (1999). Moran, E.T. Jr (1999). Nutritional effects on meat flavour. In: *Poultry Meat Science, Poultry Science Symposium Series Volume 25*, Edited by Richardson, R.I. and Mead, G.C., CABI Publishing, Wallingford, UK, pp. 197-216

Farmer, L.J., Perry, G.C., Lewis, P.D., Nute, G.R., Piggot, J.R. and Patterson, R.L.S. (1997). Responses of two genotypes of chicken to the diets and stocking densities of conventional and Label Rouge Production systems - II Sensory aspects. *Meat Science* **47**: 77-93

Farmer, L.J. (1999). Poultry meat flavour. In: *Poultry Meat Science, Poultry Science Symposium Series Volume 25,* Edited by Richardson, R.I. and Mead, G.C., CABI Publishing, Wallingford, UK, pp. 127-158

Fletcher, D.L. (1999a). Poultry meat colour. In: *Poultry Meat Science, Poultry Science Symposium Series Volume 25*, Edited by Richardson, R.I. and Mead, G.C., CABI Publishing, Wallingford, UK, pp. 159-176

Fletcher, D.L. (1999b). Broiler breast meat colour variation, pH and texture. *Poultry Science* **78**: 1323-1327.

Fletcher, D.L., Janky, D.M., Voitle, R.A. and Harms, R.H. (1977). The influence of light on broiler pigmentation. *Poultry Science* **56**: 953-956

Fletcher, D.L., Qiao, M. and Smith, D.P. (2000). The relationship of raw breast meat colour and pH to cooked meat colour and pH. *Poultry Science* **79**: 784-788

Foods Standards Agency (2000). Qualitative research to explore public attitudes to food safety. Report May 30th 2000, prepared by Cragg Ross Dawson Ltd, London

Fraps (1943). Relation of protein, fat and energy of the rations to the composition of chickens. *Poultry Science* **22**: 421

Freeman, B.M. (1983). Energy metabolism. In: *Physiology and biochemistry of the domestic fowl.* Edited by Freeman, B.M., Academic Press, London, UK, **4**: 137-377

Freeman, B.M., Manning, A.C.C. and Flack, I.H. (1980). Short-term stressor effects of food withdrawal on the immature fowl. *Comparative Biochemistry and Physiology* **67A**: 569-571

Freeman, B.M., Manning, A.C.C. and Flack, I.H. (1981). The effects of restricted feeding on adrenal corticol activity in the immature fowl. *British Poultry Science* **22**: 295-303

Froning, G.W. and Hartung, T.E. (1967). Effect of age, sex and strain on colour and texture of turkey meat. *Poultry Science* **46**: 1261

Fumiere, O., Sinnaeve, G. and Dardenne, P. (2000). Attempted authentication of cut pieces of chicken meat from certified production using near infrared spectroscopy. *Journal Near Infrared Spectroscopy* **8**: 27-34

Gasser, U. and Grosch, W. (1990). Primary odorants of chicken broth - a comparative study with meat broths from cow and ox. *Zeitschrift fur Lebensmittel Unterschung and Forschung* **190**: 3-8

Gleadthorpe Experimental Husbandry Farm Poultry *Booklet Number 7* (1980). Ministry of Agriculture, Fisheries and Food

Gleadthorpe Experimental Husbandry Farm Poultry *Booklet Number 9* (1982). Ministry of Agriculture, Fisheries and Food

Gleadthorpe Experimental Husbandry Farm Poultry *Booklet Number 10* (1983). Ministry of Agriculture, Fisheries and Food

Goldspink, G. (1985). Malleability of the motor system: a comparative approach. *Journal of Experimental Biology* **115**: 375-391

Goldspink, G., Scutt, A., Loughna, P.T., Wells, D.J., Jaenicke, T. and Gerlach, G.F. (1992). Gene expression in skeletal muscle in response to stretch and force generation. *American Journal of Physiology* **66**: 379-388

Goldspink, G. and Yang, S.Y. (1999). Muscle structure, development and growth. In: *Poultry Meat Science*, Poultry Science Symposium Series Volume 25, Edited by Richardson, R.I. and Mead, G.C., CABI Publishing, Wallingford, UK, pp. 3-18

Gompertz, B. (1825). On the nature of the function expressive of the law of human mortality, and on a new method of determining the value of life contingencies. *Phil. Trans. Royal Society,* **115**: 513-585

Goodwin, T.L. and Simpson, M.D. (1973). Chemical composition of broilers. *Poultry Science* **52**: 2032

Gordon, S. H., (1994). Effects of daylength and increasing daylength programmes on broiler welfare and performance. *World's Poultry Science Journal* **50**: 269-282

Gordon, S.H., Charles, D.R. and Green, G. (2001) Metabolic age: a basis for comparison of traditional breeds of meat chickens. *British Poultry Science* **42 (supplement)** 118-119.

Gordon, S.H. and Tucker, S.A. (1995). Effect of daylength on broiler welfare. *British Poultry Science* **36**: 844-845

Griffiths, G.L. and Nairn, M.E. (1984). Carcass downgrading of broiler chickens. *British Poultry Science* **25**: 441-446

Grosch, W. (1982). Lipid degradation products and flavours. In: *Food Flavours*. Edited by Morton, I.D. and MacLeod, A.J., Elsevier, Amsterdam, The Netherlands, pp. 325-398

Halverson, D.B. and Jacobson, M. (1970). Variations in the development of muscles in chicken. *Poultry Science* **49**: 132-136

Halverson, J.C., Waibel, P.E., Oju, E.M., Moll, S.L. and El Halwani, M.E. (1991). Effect of diet and population density on male turkeys under various environmental conditions. 2. Body composition and meat yield. *Poultry Science* **70**: 936-940

Hamdy, M.K., May, K.N. and Powers, J.J. (1961). Some physical and physiological factors affecting bruising. *Poultry Science* **40**: 790-795

Hamm, R. (1982). Post mortem changes in muscle with regard to processing of hot-boned beef. *Food Technology* **36**(11): 105-115

Hammond, J. (1932). Pigs for pork and pigs for bacon. *Journal of Agricultural Society of England* **93**: 131-145

Hammond, J. (1960). *Farm animals: their breeding, growth and inheritance.* Third edition, Edward Arnold Publishers, London, UK

Hardy, T. (1872). *Under the Greenwood Tree.* Edition (1994), Wordsworth Editions Ltd, Hertfordshire, UK

Hargis, P.S and van Elswyk, M.E. (1993). Manipulating the fatty acid composition of poultry meat and eggs. *World's Poultry Science Journal* **49**: 251-264

Havenstein, G.B., Ferket, P.R., Scheideler, S.E. and Larson, B.T. (1994). Growth, liveability and feed conversion efficiency of 1957 vs 1991 broilers when fed "typical" 1957 and 1991 broiler diets. *Poultry Science* **73**: 1785-1794

Hay, J.D., Currie, R.W., Wolfe, F.H. and Sanders, E.J. (1973). Effect of postmortem aging on chicken muscle fibrils. *Journal of Food Science* **38**: 981-986

Herring, H.K., Cassens, R.G. and Briskey, E.J. (1965). Further studies on bovine muscle tenderness as influenced by carcass position, sarcomere length, and fibre diameter. *Journal of Food Science* **30**: 1049-1054

Holland, B., Welch, A.A., Unwin, I.D., Buss, D.H., Paul, A.A. and Southgate, D.A.T. (1991) McCance and Widdowson's The composition of foods. Royal Society of Chemistry and Ministry of Agriculture, Fisheries and Food. 5th Edition Reprinted 1995

Howlinder, M.A.R. and Rose, S.P. (1989). Rearing temperature and the meat yield of broilers. *British Poultry Science* **30**: 61-67

Huges, D. and Ray, D. (1999). Developments in the global food industry. A twenty first century view. Wye College, London

Hulan, H.W., Ackman, R.G., Ratnayke, W.M.N. and Proudfoot, F.G. (1988). Omega-3 fatty acid levels and performance of broiler chickens fed redfish meal or redfish oil. *Canadian Journal of Animal Science* **68**: 533-547

Hurwitz, S., Weiselberg, M., Eisner, U., Bartov, I., Riessenfeld, G., Sharvit, M., Nair, A. and Bornstein, S. (1980). The energy requirements and performance of growing chickens and turkeys as affected by environmental temperature. *Poultry Science* **59**: 2290-2299

Huxley, H.E. (1969). The mechanism of muscular contraction. *Science* **164**: 1356-1365

Iwamato, H., Hara, Y. Gotoh, T.Y., and Takahara, H. (1993a). Different growth rates of male chicken skeletal muscles related to their histochemical properties. *British Poultry Science* **34**: 925-938

Iwamato, H., Hara, Y. Gotoh, T.Y., and Takahara, H. (1993b). Breed differences in the histochemical properties of the *M. pubo-ischio-femoralis pars medialis* myofibre of domestic cocks. *British Poultry Science* **34**: 309-322

Iwamato, H., Morita, S., Ono, Y., Takahara, H., Higashiuwatoko, H., Kukimoto, T. and Gotoh, S. (1984). A study of the fibre composition of breast and thigh muscles in Satsumadori crossbred broilers. *Japanese Journal of Zootechnical Science* **55**: 87-94

Iwamato, H. and Takahara, H. (1971). Fundamental studies on the meat production of the domestic fowl, II. Postnatal growth of the skeletal muscle in comparison with that of several other tissues and its sexual differences. *Science Bulletin of the Faculty of Agriculture, Kyushu University* **25**: 163-172

Iwamato, H., Takahara, H. and Okamato, M. (1975). Fundamental studies on the meat production of the domestic fowl, VI. Postnatal growth of skeletal muscle, skin, viscera, bone and fatty tissue of Barred Plymouth Rock chicken. *Science Bulletin of the Faculty of Agriculture, Kyushu University* **29**: 151-162

Janky, D.M., Fletcher, D.L., Voitle, R.A. and Harms, R.H. (1980). The influence of light transmission properties of plastic window coverings on broiler pigmentation. *Poultry Science* **59**: 1350-1352

Jankey, D.M., Voitle, R.A. and Harms, R.H. (1985). The influence of different xanthophyll-containing feedstuffs on pigmentation of broilers reared in open and windowless houses. *Poultry Science* **64**: 925-931

Jonnson, G. and McNab, J.M. (1983). Grass meal as an ingredient in diets for broiler chickens. *British Poultry Science* **24**: 361-369

Kafri, I., Cherry, J.A., Jones, D.E. and Siegel, P.B. (1985). Breaking strength and composition of the skin of broiler chicks: response to dietary calorie-protein ratios. *Poultry Science* **64**: 2143-2149

Kafri, I., Jortner, B.S. and Cherry, J.A. (1986). Skin breaking strength in broilers: relationship with skin thickness. *Poultry Science* **65**: 971-978

Kang, C.W., Sunde, M.L. and Swick, R.W. (1985). Growth and protein turnover in the skeletal muscles of broiler chicks. *Poultry Science* **64**: 370-379

Karunajeewa, H. (1987). A review of current poultry feeding systems and their potential acceptability to animal welfarists. *World's Poultry Science Journal* **43**: 20-32

Katanbaf, M.N., Dunnington, E.A. and Siegel, P.B. (1989). Restricted feeding in early and late-feathering chickens. *Poultry Science* **68**: 359-368

Keeton, J.T. (1993). Low-fat meat products - technological problems with processing. *Meat Science* **36**: 261-276

Khan, A.W. (1975). Effect of chemical treatments causing rapid onset of rigor on tenderness of poultry breast meat. *Journal of Agricultural Food Chemistry* **23**: 449-451

Kijowski, J., Niewiarowicz, A. and Kujawska-Biernat, B. (1982). Biochemical and technological characteristics of hot chicken meat. *Journal of Food Technology* **17**: 553-560

King, A.J., Uijttenboogaart, T.G. and de Vries, A.W. (1995). a-tocopherol, b carotene and ascorbic acid as antioxidants in stored poultry muscle. *Journal of Food Science* **60**: 1009-1012

Knizetova, H., Hyanek, J., Knize, B. and Roubicek, J. (1991). Analysis of growth curves of fowl. I. Chickens. *British Poultry Science* **32**: 1027-1038

Kranen, R.W., Scheele, C.W., Veerkamp, C.H., Lambooy, E., Kuppevelt, T.H. and Veerkamp, J.H. (1998). Susceptibility of broiler chickens to haemorrhages in muscles: the effect of stock and rearing temperature regimen. *Poultry Science* **77**: 334-341

Kranen, R.W., Lambooy, E., Veerkamp, C.H., Kuppevelt, T.H., Veerkamp, J.H. (1999). Histological characterisation of hemorrhages in muscles of broiler chicks. *Poultry Science* **78**: 467-476

Laird, A.K. (1966). Postnatal growth of birds and mammals. *Growth* **30**: 349-363

Larbier, M. and Leclercq, B. (1992). *Nutrition and Feeding of Poultry*. Translated and edited Wiseman, J., Nottingham University Press, Nottingham, UK

Laszczyk-Legendre, A. (1999). Label Rouge traditional free range poultry: a concept including quality, environment and welfare. In: *Proceedings from the 14th European Symposium on the Quality of Poultry Meat*, World's Poultry Science Association, Bolongna, pp. 255-263

Lawrence, T.L.J. and Fowler, V. (1997). *Growth of farm animals.* CAB International, Wallingford, UK

Lawrie, R.A. (1992). *Meat Science.* Fifth Edition, Pergamon, Oxford, UK

Lewis, P.D., Perry, G.C., Farmer, L.J. and Patterson, R.L.S. (1997). Responses of two genotypes of chicken to the diets and stocking densities typical of UK and 'Label Rouge' Production systems: I Performance, behaviour and carcass composition. *Meat Science* **45**: 501-516

Light, N.D., Champion, A.E., Voyle, C. and Bailey, A.J. (1985). The role of epimysial collagen in determining texture in six bovine muscles. *Meat Science* **13**: 137-149

Lin, C.F., Gray, J.I., Asghar, A., Buckley, D.J., Booren, A.F. and Flegel, C.J. (1989). Effects of dietary oils and a-tocopherol supplementation on lipid composition and stability of broiler meat. *Journal of Food Science* **54**: 1457-1460, 1484

Lipstein, B., Bornstein, S. and Bartov, I.(1975). The replacement of some of the soyabean meal by the first-limiting amino acids in practical broiler diets. 3. Effects of protein concentrations and amino acid supplementations in broiler finisher diets on fat deposition in the carcass. *British Poultry Science* **16**: 627-635

Locker, R.H. (1960). Degree of muscular contraction as a factor in tenderness of beef. *Food Research* **25**: 304-307

Lynn, N.J., Tucker, S.A. and Bray, T.S. (1991). Litter condition and contact dermatitis in broiler chickens. In: *Quality of poultry products poultry meat. Proceedings of Spelderholt Jubilee Symposia,* Doorwerth, Netherlands, May 1991. Edited by Vijttenboogart, T.G. and Veerkamp, C.H.

Lyon, C.E. and Buhr, R.J. (1999). Biochemical basis of meat texture. In: *Poultry Meat Science, Poultry Science Symposium Series Volume 25,* Edited by Richardson, R.I. and Mead, G.C., CABI Publishing, Wallingford, UK, pp. 99-126

Lyon, C.E., Hamm, D., Hudspeth, J.P. and Benoff, F.H. (1984). Effects of age and sex on the texture profile of hot stripped broiler breast meat. *Poultry Science* **63**: 2508-2510

McCarthy, J.C. (1977). Quantitative aspects of the genetics of growth. In: *Growth and Poultry Meat Production,* Edited by Boorman, K.N. and Wilson, B.J., British Poultry Science Ltd, Edinburgh, UK, pp. 117-130

McCormick, R.J. (1999). Extracellular modifications to muscle collagen: implications for meat quality. *Poultry Science* **78**: 785-791

MacDonald, P., Edwards, R.A. and Greenhalgh, J.F.D. (1973). *Animal nutrition*. Second Edition. Longman, New York, USA

MacDonald, M.L. and Swick, R.W. (1981). The effect of protein depletion and repletion on muscle-protein turnover in the chick. *Biochemical Journal* **194**: 811-819

MacFarland, D.C. (1999). Influence of growth factors on poultry myogenic satellite cells. *Poultry Science* **78**: 747-758

MacLeod, M.G., McNiel, L., Knox, A.I. and Bernard, K. (1998). Comparison of bodyweight responses to dietary lysine concentration in broilers of two commercial lines and a "relaxed-selection" line. In: *Proceedings of the World's Poultry Science Association (UK Branch) Spring Meeting*, Scarborough, pp. 41-42

McIlroy, S.G., Goodall, E.A. and McMurray, C.H. (1987). A contact dermatitis of broilers - epidemilogical findings. *Avian Pathology* **16**: 93-105

McMeekan, C.P. (1940). Growth and development of the pig, with special reference to carcass quality characters. *Journal of Agricultural Science* **30**: 387-436.

Maga, J.A. (1994). Pink discolouration in cooked white meat. *Food Reviews International* **10**(3): 273-286

Marsh, B.B. (1954). Rigor mortis in beef. *Journal of the Science of Food and Agriculture* **5**: 70-75

Maruyama, K., Sunde, M.L. and Swick, R.W. (1978). Growth and muscle protein turnover in the chick. *Biochemical Journal* **176**: 573-582

Mead, G.C., Griffiths, N.M., Impey, C.S. and Coplestone, J.C. (1983). Influence of diet on the intestinal microflora and meat flavour of intensively-reared broiler chickens. *British Poultry Science* **24**: 261-272

Melton, S.L. (1990). Effects of feeds on flavour of red meat: a review. *Journal of Animal Science* **68**: 4421-4435

Meltzer, A. (1983a). The effect of body temperature on the growth rate of broilers. *British Poultry Science* **24**: 471-476

Meltzer, A. (1983b). Thermoneutral zone and resting metabolic rate of broilers. *British Poultry Science* **24**: 489-495

Meltzer, A. (1987). Acclimatisation to ambient temperature and its nutritional consequences. *World's Poultry Science Journal* **43**: 33-44

Merit, E.S. (1974). Selection for growth rate in broilers with a minimum increase in adult size. In: *Proceedings of 1st World Congress of Genetic Applied Livestock Production*, Madrid **1**: 951-958

Millares, R. and Fellers, C.R. (1948). Amino acid content of chicken. *Journal of American Dietetics Association* **24**: 1057

Miller, R.K. (1994). Quality characteristics. In: *Muscle Foods; Meat, Poultry and Seafood Technology*. Edited by Kinsman, D.M., Kotula, A.W. and Breidenstein, B.C., Chapman & Hall, New York, USA, pp. 296-332

Ministry of Agriculture, Fisheries and Food (1999) National Food Survey 1998. The Stationery Office, London

Ministry of Agriculture, Fisheries and Food (August 2000). Poultry and Poultrymeat Statistics Notice.

Moore, G.E. and Goldspink, G. (1985). The effect of reduced activity on the enzymatic development of phasic and tonic muscles in the chicken. *Journal of Developmental Physiology* **7**: 381-386

Moran, E.T. Jr. (1994). Response of broiler strains differing in body fat to inadequate methionine, live performance and processing yields. *Poultry Science* **73**: 1116-1126

Moran, E.T., Jr (1996). Broiler feeding regimen and yield. In: *Proceedings of the Western Canada Nutrition Conference*, Edmonton, Canada, pp. 5-13

Moran, E.T. Jr. (1999). Live production factors influencing yield and quality. In: *Poultry Meat Science, Poultry Science Symposium Series Volume 25*, Edited by Richardson, R.I. and Mead, G.C., CABI Publishing, Wallingford, UK, pp. 179-195

Moran, E.T. Jr. and Bilgili, S.F. (1989). Processing losses, carcass quality and meat yield of broiler chickens receiving diets marginally deficient to adequate in lysine prior to marketing. *Poultry Science* **69**: 702-710

Moran, E.T., Jr and Stilborn, H.L. (1996). Effect of glutamic acid on broilers given submarginal crude protein and adequate essential amino acids using feeds high and low in potassium. *Poultry Science* **75**: 120-129

National Research Council (1994). *Nutrient requirements for poultry*. Ninth revised edition. National Academy Press, Washington, USA

Newbold, R.P. (1966). Changes with associated rigor mortis. In: *The physiology and biochemistry of muscle as a food*. Edited by Briskey, E.J., Cassens, R.G. and Trautman, J.C., University of Wisconsin Press, Madison, USA, pp. 213-224

Newton, K.G. and Gill, C.O. (1981). The microbiology of DFD fresh meats. A review. *Meat Science* **5**: 223-232

Nir, I., Nitzan, Z. and Mahagna, M. (1993). Comparative growth and development of digestive organs and of some enzymes in broiler and egg type chicks after hatching. *British Poultry Science* **34**: 523-532

Northcutt, J.K, Buhr, R.J. and Rowland, G.N. (2000). Relationship between broiler bruise age to appearance and tissue histological characteristics. *Applied Poultry Research* **9**: 13-20

Olomu, J.M. and Baracos, V.E. (1991). Influence of dietary flaxseed oil on the performance, muscle protein deposition and fatty acid composition of broiler chickens. *Poultry Science* **70**: 1403-1411

Ono, Y., Iwamato, H. and Takahara, H. (1982). Studies on the growth of the skeletal muscle of capon. II. Effects of castration on muscle weights in different body parts and individual muscle weights. *Science Bulletin of the Faculty of Agriculture, Kyushu University* **37**: 20-30

Ono, Y., Iwamato, H. and Takahara, H. (1983). Histochemical studies on the effects of androgen on the fibre composition of M. biceps femoris in cocks. *Japanese Journal of Zootechnical Science* **54**: 453-459

Orr, H.L., Hunt, E.C. and Randall, C.J. (1984). Yield of carcass, parts, meat, skin and bone of eight strains of broilers. *Poultry Science* **63**: 2197-2200

Paul, A.A., Southgate, D.A.T. and Russell, J. (1980) First supplement to McCance and Widdowson's The composition of foods. HMSO, London, UK

Pingel, H. and Knust, U. (1993). Review on duck meat quality. In: *Proceedings of the XI European Symposium on the Quality of Poultry Meat*. Tours, France, **1**: 26-38

Pooni, G.S. and Mead, G.C. (1984). Prospective use of temperature function integration for predicting the shelf-life of non-frozen poultry-meat products. *Food Microbiology* **1**: 67-78

Prescott, N.J., Wathes, C.M., Kirkwood, J.K. and Perry, G.C. (1985). Growth, food intake and development in broiler cockerels reared to maturity. *Animal Production* **41**: 239-245

Raikov, R.J. (1985). Locomotor system. In: *Form and Function in Birds*. Edited by King, A.S. and McLelland, J., London Academic Press **3**: 57-147

Remignon, H., Gardahaut, M.-F., Marche, G. and Richard, F.-H. (1995).

Selection for rapid growth increases the number and size of muscle fibres without changing their typing in chickens. *Journal of Muscle Research and Cell Motility* **16**: 95-102

Remignon, H., Lefaucheur, L., Blum, J.C. and Ricard, F.H. (1994). Effects of divergent selection for body weight on three skeletal muscle characteristics in the chicken. *British Poultry Science* **35**: 65-76

Renden, J.A. Bilgili, S.F. and Kincaid, S.A.(1992). Effects of photoschedule and strain cross on broiler performance and carcass yield. *Poultry Science* **71**: 1417-1426

Renden, J.A., Bilgili, S.F. and Kincaid, S.A.(1993). Research note: Comparison of restricted and increasing light programs for male broiler performance and carcass yield. *Poultry Science* **72**: 378-382

Ricard, F.H. and Touraille, C. (1988). Influence du sexe sur les characteristiques organoleptiques de la viande de poulet. *Archives Geflugelkunde* 52: 27-30

Ricklefs, R.E. (1985). Modification of growth and development of muscles of poultry. *Poultry Science* **64**: 1563-1576

Roberts, V. (1997). *British Poultry Standards*, Fifth Edition, Blackwell Science, Oxford, UK

Robinson, L. (1948). *Modern Poultry Husbandry*. Crosby and Lockwood, London, UK

Rose, S.P. (1997). *Principles of poultry science*. CAB International, Wallingford, UK

Roth, F.X., Kirchgessner, M., Ristic, M., Kreuzer, M. and Mansrus-Kuhral, E. (1990). Amino acid pattern of the breast meat of broilers during an extended finishing period as affected by protein and energy intake. *Fleischwirtschaft* **70**: 608-612

Sams, A.R. (1999). Meat quality during processing. *Poultry Science* **78**: 798-803

Sandusky and Heath (1988a). Effect of age, sex and barriers in experimental pens on muscle growth. *Poultry Science* **67**: 1708-1716

Sandusky, C.L. and Heath, J.L. (1988b). Growth characteristics of selected broiler muscles as affected by age and experimental pen design. *Poultry Science* **67**: 1557-1567

Savory, J. (1976). Broiler growth and feeding behaviour in three different lighting regimes. *British Poultry Science* **17**: 557-560

Scheurs, F.J.G., van der Heide, D., Leenstra, F.R. and de Wit, W. (1995). Endogenous proteolytic enzymes in chicken muscles. Differences among strains with different growth rates and protein efficiencies. *Poultry Science* **74**: 523-527

Scholtyseek, S. and Sailor, K. (1986). Differences in the taste of poultry meat. *Archives Geflugelkunde* **50**: 49-54

Shrimpton, D.H. and Miller, W.S. (1960). Some causes of toughness in broilers (young roasting chickens). II. Effects of breed, management and sex. *British Poultry Science* **1**: 111-120

Sibbald, I.R. and Wornetz, M.S. (1986). Effects of dietary lysine and feed intake on energy utilisation and tissue synthesis by broiler chicks. *Poultry Science* **65**: 98-105

Simpson, M.D. and Goodwin, T.L. (1979). Chemical composition and yield of Cornish Game hens and broilers. *Poultry Science* **58**: 1400-1402

Singh, S.P. and Essary, E.O. (1971). Vitamin content of broiler meat as affected by age, sex, thawing and cooking. *Poultry Science* **50**: 1150-1155

Skaarup, T. (1983). The quality of meat from free range chickens versus chickens in confinement. In: *Proceedings of European Symposium on Quality of Poultry Meat.* Edited by Lahelle, C., Ricard, F.H. and Colin, P., Ministère de l'Agriculture, Station Expérimentale d'Aviculture, Ploufragon, France, pp. 37-45

Smith, D. (1999). *How to Cook Book Two*. BBC Worldwide Ltd, London, UK

Smith, D.P. and Fletcher, D.L. (1988). Chicken breast muscle fibre type and diameter as influenced by age and intramuscular location. *Poultry Science* **67**: 908-913

Smith, J.H. (1963). Relation of body size to muscle cell size and number in the chicken. *Poultry Science* **42**: 283-290

Smith, S.B. (1991). Dietary modification for altering fat composition of meat. In: *Fat and Cholesterol Reduced Foods: Technologies and Strategies.* Edited by Harberstroh, C. and Morris, C.E., the Portfolio Publishing Company, The Woodlands, Texas, USA, pp. 75-97

Smith, M.O. (1993). Parts yields of broilers reared under cycling high temperatures. *Poultry Science* **72**: 1146-1150

Smith, M.O. and Teeter, R.G. (1987). Influence of feed intake and ambient temperature stress on the relative yield of broiler parts. *Nutritional Reports International* **35**: 299-306

Smith, T.W., Jr., Couch, J.R., Garret, R.L. and Creger, C.R. (1977). The effect of sex, dietary energy, meat protein, ascorbic acid and iron on broiler skin collagen. *Poultry Science* **56**: 1216-1220

Snow, M.H. (1977). The effects of aging on satellite cells in skeletal muscles of mice and rats. *Cell and Tissue Research* **185**: 399-408

Sonaiya, E.B., Ristic, M. and Klein, F.W. (1990). Effect of environmental temperatures, dietary energy, age and sex on broiler carcass portions and palatability. *British Poultry Science* **31**: 121-128

Spearman, R.J.C. (1971). Integumentary system. In: *Physiology and Biochemistry of the Domestic Fowl. Volume 2.* Edited by Bell, D.J. and Freeman, B.H., Academic Press, London, UK, pp. 603-619

Streiff, K., Volker, L. and Friesecke, H. (1977) Dietary vitamin E and rancidity of poultry tissue. In: *Growth and poultry meat production.* Edited by Boorman, K.N. and Wilson, B.J., Poultry Science Symposium No.12, British Poultry Science Ltd., pp. 335-340

Summers, J.D. and Leeson, S. (1985). Broiler carcass composition as affected by amino acid supplementation. *Canadian Journal of Animal Science* **65**: 717-723

Summers, J.D., Leeson, S. and Spratt, D. (1988). Yield and composition of edible meat from male broilers as influenced by dietary protein level and amino acid supplementation. *Canadian Journal of Animal Science* **68**: 241-248

Susbilla, J.P., Frankel, T.L., Parkinson, G. and Gow, C.B. (1994). Weight of internal organs and carcass yield of early food restricted broilers. *British Poultry Science* **35**: 677-685

Sutcliffe, N.A., King, A.W.M. and Charles, D.R. (1987) Monitoring poultry house environment. In: *Computer applications in agricultural environments.* Edited by Clark, J.A., Gregson, K. and Saffell, R.A., Butterworths, London, UK, pp. 207-218

Taylor, C.S. (1965). A relationship between mature weight and time taken to mature in mammals. *Animal Production* **7**: 203-20

Taylor, M.H. and Helbecka, N.V.L. (1968). Field based studies of bruised poultry. *Poultry Science* **47**: 1166-1169

Telford, M.E., Holroyd, P. and Wells, R.W. (1986). *History of the National Institute of Poultry Husbandry*, Nuffield Press

Tesseraud, S., Maaa, N., Peresson, R. and Chagneau, A.M. (1996). Relative responses of protein turnover in three different skeletal muscles to dietary lysine deficiency in chicks. *British Poultry Science* **37**: 641-650

Thompson, A. (1952). *The complete poultryman.* Faber and Faber, London, UK

Touraille, C., Ricard, F.H., Kopp, J., Valin, C. and Leclercq, B. (1981). Chicken meat quality. 2. Changes with age of some physio-chemical and sensory characteristics of meat. *Archives Geflugelkunde* **45**: 97-104

Tucker, S.A. and Walker, A.W. (1992). Hock burn in broilers. In: *Recent Advances in Animal Nutrition*, Edited by Garnsworthy, P.C., Haresign, W. and Cole, D.J.A., Butterworth and Heinemann, Oxford, UK, pp. 33-50

Tzeng, R.Y. and Becker, W.A. (1981). Growth patterns of body and abdominal fat weights in male broilers. *Poultry Science* **60**: 1101-1106

Van Kampen, (1981). Thermal influence on poultry. In: *Environmental aspects of housing for animal production.* Edited by Clarke, J.A., Butterworth, London, UK, pp. 131-147

Van Kampen, G.J.M. and Jansman, A.J.M. (1994). Use of EC produced oil seeds in animal feeds. In: *Recent Advances in Animal Nutrition.* Edited by Garnsworthy, P.C. and Cole, D.J.A., Nottingham University Press, Nottingham, UK, pp. 31-56

Velleman, S.G., Yeager, J.D., Krider, H., Carrino, D.A., Zimmerman, S.D. and McCormick, R.J. (1996). The effect of proteoglycans on the morphology of collagen fibrils formed in vitro. *Collagen Related Research* **7**: 105-114

Vollmerhaus, B. (1992). Spezielle anatomie des bewegungsapparatus. In: *Lehrbuch der Anatomie de haustiere, Band, V. Anatomie der Vogel.* Edited by Schummer, A. and Vollmerhause, B., Paul Parey, Berlin, Germany, pp. 54-154

Waite, R. and Sastry, K.N.S. (1949). *Empirical Journal of Experimental Agriculture* **17**: 179

Walker, A.W., Wiseman, J., Lynn, N.J. and Charles, D.R. (1995). Recent findings on the effects of nutrition on the growth of specific broiler carcass components. In: *Recent Advances in Animal Nutrition*, Edited by Haresign, W. and Cole, D.J.A., Nottingham University Press, Nottingham, UK, pp. 169-184

Warren, D.C. (1958). A half century of advances in the genetics and breeding improvement of poultry. *Poultry Science* **37**: 3-20

Warris, P.D., Wilkins, L.J. and Knowles, T.G. (1999). The influence of ante-mortem handling on poultry meat quality. In: *Poultry Meat Science, Poultry Science Symposium Series Volume 25*, Edited by Richardson, R.I. and Mead, G.C., CABI Publishing, Wallingford, UK, pp. 217-230

Wathes, C.M., Gill, B.D., Charles, D.R. and Back, H.L. (1981). The effects of temperature on broilers: a simulation model of the responses to temperature. *British Poultry Science* **22**: 483-492

Wheeler, K.B. and Latshaw, J.D. (1981). Sulphur amino acid requirements and interactions in broilers during two growth periods. *Poultry Science* **60**: 228-236

Whittle, T.E. (1998). A triumph of science. A 70 year history of the UK poultry industry. *Poultry World*

Wick, M. (1999). Filament assembly properties of the sarcomeric myosin heavy chain. *Poultry Science* **78**: 735-742

Wilkins, L.J., Brown, S.N., Phillips, A.J. and Warriss, P.D. (2000). Variation in the colour of broiler breast fillets in the UK. *British Poultry Science* **41**: 308-312

Wilson, B.J. (1977). Growth curves: their analysis and use. In: *Growth and Poultry Meat Production*. Edited by Boorman, K.N. and Wilson, B.J., British Poultry Science Ltd, Edinburgh, UK, pp. 89-115

Winsor, C.P. (1932) The Gompertz curve as a growth curve. *Proceedings of the National Academy of Science* **18**: 1-8

Wladyka, E.J. and Dawson, L.E. (1968). Proximate composition of thawed chicken meat and drip after storage. *Poultry Science* **47**: 1111

Yahov, S., Straschinow, A., Plavnik, I. and Hurwitz, S. (1996). Effects of diurnally cycling versus constant temperatures on chicken growth and food intake. *British Poultry Science* **37**: 43-54

Yamashita, C., Ishimoto, Y., Mekada, H., Ebisawa, S., Murai, T. and Nonaka, S. (1976). *Japanese Journal of Poultry Science* **13**: 14-19

Yan, J.C., Denton, J.H., Bailey, C.A. and Sams, A.R. (1991). Customising the fatty acid content of broiler tissues. *Poultry Science* **70**: 167-172

7

PASTURE TYPES AND PASTURE MANAGEMENT

Traditional pastures

In the traditional period, authors such as Robinson (1948) stressed the need for grasses of high nutritional value to be included in the sward. The feeding value of grass to poultry was taken seriously. Robinson (1948) recommended keeping the grass short, and quoted estimates of feed savings of up to 10% or more from the grazing of high quality young fresh grass, although the contribution was debated. Thompson (1952) considered that properly managed, short, good quality pasture could be so beneficial in spring and early summer as to reduce the consumption of feed by 5%, though he described "out of hand" pasture as useless. Grass was also considered to contribute to an acceptable yolk colour.

In a handbook of guidelines Poultry World (1959) mentioned that many different mixtures had been tried, but gave an example sward composition (Table 1). The sward was dominated by perennial ryegrass (*Lolium perenne*), though an alternative based on Timothy (*Phleum pratense*) was offered.

Table 1. A TYPICAL TRADITIONAL SEEDS MIXTURE FOR POULTRY PASTURE

Species	*Quantity (lb)*
Perennial ryegrass (Aberystwyth 23)	20
Creeping red fescue	3
Wild white clover (English certified)	1
White clover (Aberystwyth 100)	1
Total	25 lb/acre (28.1 kg/ha)

Source: Poultry World (1959)

In the last 15 to 20 years of the conventional period, there was a revival of non-organic free-range egg production. This was due to consumer demands for non-cage eggs. The consumers' perception of free-range egg production was important. Free range eggs are a premium product and the extra costs of production will only be met by the consumer if the system meets expectations. Thus, there was a need for the pasture to be hard wearing and resilient to poaching. Poached muddy areas close to the house are likely to be viewed unfavourably by the consumer, and when this occurs management of the house litter is more difficult. This meant that swards were mostly composed of perennial ryegrass.

The seed mixture used in DEFRA-funded project OF0153 is typical and is given in Table 2. It proved to be hard wearing for meat birds during a wet summer, and the birds were observed to eat it.

Table 2. SEEDS MIXTURE USED IN PROJECT OF0153

Species	*Quantity (kg)*
Amenity perennial ryegrass	31.50
Smooth stalked meadow grass	4.50
Brown top bent	2.25
Timothy	2.25
White clover	4.50
Total	45 kg/ha

Source: DEFRA-funded project OF0153

Future needs

In the conventional period the contribution of grazing to the nutrition of the birds has usually been considered to be negligible. This approach is precautionary: nutrients for maintenance and optimal performance are included in the feed so as to reduce the risk of poor performance or "searching" behaviour. Searching occurs when some nutrients are low in the ration. For example, low feed methionine content is thought by many poultry professionals, from field observations, to give rise to feather pecking (e.g. Owen, 2002, *personal communication*). Feathers are a rich source of sulphur containing amino acids (Larbier and Leclercq, 1994, quoting Boorman and Burgess, 1986). However, in free-range situations it is more difficult to accurately predict feed intake because of the influence of temperature on energy requirements. Although the ratio of metabolisable energy value to protein content of the ration is usually altered so as to take into account seasonal effects of temperature on intake, there is likely to be an over consumption of feed protein during cold weather and the converse during hot weather. This means that it would be nutritionally useful to be able to rely on the pasture acting as a buffer for energy or protein.

In organic systems of poultry production the pasture should contribute to the birds' nutrient requirements and manuring of the pasture recycles nutrients to the land: this is part of the ethos of the systems. Furthermore, if poultry were to forage high nutrient value swards then this would reduce the reliance on expensive bought-in feeds and it would improve the sustainability of the system. The use of imported proteinaceous ingredients in organic poultry feeds greatly reduces sustainability.

The standards for organic livestock systems promote the use of home-grown feed for livestock and require that 50% of the diet fed is certified organic within poultry systems (UKROFS, 2001). This means that organic poultry producers are currently seeking alternative methods of providing high quality home-grown organic feed to poultry. Some alternative forages contain high levels of protein and energy, and in particular, high concentrations of essential amino acids.

Lastly, there may be the opportunity to include herbage having antioxidant or antibiotic activity within the pasture, such as rosemary, sage or oregano (Adams, 1999). If consumed this may benefit bird immunocompetence and performance. Herb intake may also affect egg or meat flavour and thus potentially provide some scope for product differentiation.

Another possible benefit of moving away from ryegrass based swards is that some forage crops may provide aerial cover for chickens. Chickens in extensive systems are reluctant to range freely on open grass swards (DEFRA-funded project OF0153), and areas of pasture close to the house become poached. Good pasture usage will be essential if birds are to derive significant nutrient intakes from the forage.

There is a need to examine the extent to which alternative forages may be used to provide home-grown feeds and cover to table birds managed in extensive poultry systems. There should also be a focus on developing innovative management systems for maximising the utilisation of pasture by extensively managed poultry.

Mixed grazing free-range layers with sheep was considered useful until about 10 to 15 years ago, but the practice has fallen out of favour for reasons of food safety. The sheep kept the grass short and they may have assisted in getting the birds to range more evenly.

In the conventional period, pasture management has been limited: the grass is mowed short and the clippings are collected so as to avoid crop impaction. Reseeding is done in poached areas so as to take advantage of spring and early summer growth. Pasture rest allows the sward to recover or establish and this is best done by paddock rotation. Maurer *et al.,* (2000), working in Switzerland, recommended paddock rotation in the interests of sward quality. Paddock rotation permits long rest periods and this may be of benefit in reducing the parasite burden on the land, although some parasites are relatively resilient to environmental conditions outside the host.

Bassler *et al.,* (1999) suggested that, without scientific studies to refer to, two or three years between land use is a provisional recommendation. This can be achieved when poultry occupy a grass/clover ley integrated into the rotation. On permanent pasture, the authors suggested that a dense sward can carry between 50 to 100 hens for seven days without the birds damaging the sward near to the house (stocking rate 15 hens per ha per year). Damage to the sward was thought to be mostly due to scratching. They recommended that the house is then moved 20 metres. The distance reflected their observations that birds remain close to the house. For bigger flocks the houses need to be moved more frequently so as to avoid damage to the sward.

Bassler *et al.,* (1999) calculated that for the above scenario of paddock rotation (i.e. a fresh 20 m radius of pasture every seven days), and when having six months vegetative cover and a two seasons rest period, a flock of 50 hens would require about 10 hectares of land over three years (3.3 ha per year, 15 hens per hectare per year). They commented that this system seemed to be extremely extensive but noted that two thirds of the "required area" may be used for other crops or animals in the meantime.

At the end of the stocking period chain harrowing would probably be advisable in order to spread the droppings. Harrowing was recommended by the authorities of the traditional period as part of parasite control programmes (e.g. Thompson, 1952).

Soil pH may be relevant to the control of parasites. The writers of the traditional period, such as Robinson (1948), stressed the need for liming to prevent the pasture becoming what was described as fowl sick.

Poultry pastures need to be well drained in the interests of grassland management, disease control and house litter management. Heavy, poorly drained soils are therefore generally unsuitable, although in fixed housing systems drainage may be used to improve the pasture. In FRFR systems good natural pasture drainage will be important.

Land topography is important as it will be difficult to stabilise mobile housing on hilly or undulating land. For laying hens the problem is likely to be exacerbated as wind may rush under the house and up through the droppings pit. In extreme weather conditions the house may even be lifted by wind.

In commercial free-range systems, hedges and trees are often thought to encourage chickens away from the house, and some thought should be given to incorporating these into the pasture. This may be more difficult to achieve in FRFR systems than when using fixed housing systems with dedicated pasture. In FRFR systems hedges and trees may interfere with ploughing, sowing and harvesting tasks, and shade may affect crop growth. However, a recent publication on Label Rouge production in France suggests that tree cover is attractive to outdoor chickens (Lubac and Mirabito, 2001), and furthermore, bird activity seems to be related to plant canopy at the visited area (Mirabito and Lubac, 2001). In heavily shaded areas, as found under established trees, birds tended to sit, whereas standing was the predominant activity in non-shaded areas. Time spent outdoors is likely to be greater if birds have what they perceive to be a safe area where resting may occur.

Complementary work by Mirabito *et al.,* (2001) looked at the effects of housing chickens in peach tree orchards on range usage. The spatial distribution of peach trees was homogeneous throughout the paddock. For comparison, a paddock without trees and ryegrass sward was used. Although there were more chickens outdoors in the orchards than in the bare paddocks, the peach trees were not successful in encouraging birds to range away from the house. This was surprising to the authors and they commented that it was not consistent with observations made in similar commercial systems.

It is likely that house design, paddock design, sward composition and cover, and bird management all influence ranging activity. In addition to these factors there will also be the influence of weather on range usage.

There is a need to consider the effect of integrating chickens into cropping systems on the soil ecology. If changes in soil ecology occur through the presence of chickens in cropping systems then there may be implications for crop disease control.

Clark and Gage (1997) studied the effects of integrating chickens into nonchemical apple orchards on the abundance of soil macroinvertebrates. Two groups of organisms were studied: epigeic predators and earthworms. The basis for the study was that chickens have been found to feed on a wide range of macroinvertebrates living in the soil surface: including ground beetles (*Carabidae*), rove beetles (*Staphlinidae*), spiders (*Araneae*) and earthworms (*Lumbricidae*) (Clark and Gage, 1996). The presence of chickens in the orchard may, therefore, alter their populations through intake. In addition, the removal of vegetation and mild soil compaction which results from foraging may alter the soil surface environment, whereas manuring may alter food webs: these may change the macroinvertebrate population. The authors pointed out that ground beetles are predators of apple maggot (Clark and Gage, 1997 citing work by Allen and Hagley, 1990), and if chickens reduce the abundance of ground beetles then there may be an increase in the apple maggot population. Earthworms have a role in maintaining soil fertility (Clark and Gage, 1997 citing Berry 1994) and they can indirectly reduce plant diseases in orchards by breaking down leaf litter (Clark and Gage, 1997 citing Raw, 1962 and Kennel, 1990).

Clark and Gage (1997) found that chickens reduced the abundance of spiders (*Araneae*) and harvestmen (*Opiliones*), but they did not reduce the abundance of ground beetles (*Carabidae*) or rove beetles (*Staphylinidae*). Most of the spiders collected in the pit traps were *Lycosids*, a family with mainly diurnally active species. The authors reported that the most common harvestman, *Phalangium opilio* L. was also diurnally active. By comparison, most *Carabids* and *Staphylinids* seek shelter under the soil surface or leaf litter in the day and this may make them less likely to be eaten by foraging chickens.

The presence of chickens over a three year period had no detectable effect on earthworm abundance or on soil organic matter (Clark and Gage, 1997).

However, the authors commented that their findings were relevant to the stocking rate used, and the latter was low by commercial standards, even for organic poultry production (about 11 m^2 pasture/chicken).

In a review of management of laying hens in mobile houses Bassler *et al.,* (1999) cited work reporting the nutrient value of earthworms (Yoshida and Hoshii, 1978), grasshoppers (Sugimura, 1984) and house-fly pupae (Teotia and Miller, 1974). The metabolisable energy values of the above were potentially useful, at between 12.4 MJ/kg and 12.8 MJ/kg dry matter (DM), and the crude protein, lysine and methionine plus cystine contents were very high. For example, in earthworms the contents were 610 g crude protein/kg DM, 42 g lysine/kg DM and 12 g methionine plus cystine/kg DM. Earthworms were lower in crude protein content than grasshoppers or house-fly pupae, but earthworms were richest in lysine and only slightly lower in methionine plus cystine content than the richest source, this being grasshoppers. By comparison, Bassler *et al.,* (1999) gave crude protein, lysine and methionine plus cystine contents of 165 g/kg DM, 6.1 g/kg DM and 5.3 g/kg DM, respectively for layers concentrate when feed intakes of 110 g/hen.day are expected (citing NRC, 1984). However, the latter intake is low by free-range standards and it is probably based on caged hens.

Bassler *et al.,* (1999) cited work by Hughes and Dunn (1983) suggesting that laying hens have the capacity to consume up to between 30 g and 40 g per day of dry matter of herbage plus small animals, in addition to more than 100 g of dry matter of concentrate. What is not known is how much of the birds' nutritional needs are met by herbage and insect intake on range. Furthermore, the contribution of nutrients from range is expected to differ widely depending on the time of year, stocking rate, quality and utilisation of the range and bird performance (Bassler *et al.,* 1999).

The use of high protein forages, and management strategies aimed at enhancing the abundance of invertebrates in the soil surface, may be desirable in terms of sustainability. However, the seasonal production of organic poultry products would be undesirable from a supply and demand point of view, and so it may be necessary to adopt seasonal feeding and management practices that take into account differences in the contribution of nutrients from the range.

In summary, there is a need to identify alternative forages for use in extensive systems of chicken production. The forages should be palatable, of high nutrient value and provide aerial cover for the birds. Consequently, best management practices will need to be determined so that forage palatability, nutrient value and pasture cover are high for most of the year. Pasture cover is essential if organic poultry production in the UK is to be a year round enterprise. Bare paddock, that is soil only, is not allowed in organic production or non-organic free-range production. Lastly, paddock rotations and land management practices need to be optimised so that parasite control is achieved, and soil ecosystems thrive.

Free-range research facilities at ADAS Gleadthorpe

References

Adams, C. A. (1999). *Nutricines: Food Components in Health and Nutrition.* Nottingham University Press

Allen, W.R. and Hagley, E.A.C. (1990). Epigeal arthropods as predators of mature larvae and pupae of the apple maggot (Diptera: Tephritidae). *Environmental Entomology* **19**:309-312

Bassler, A., Ciszuk, P. and Sjelin, K. (1999). Management of laying hens in mobile houses – a review of experiences. In: *Ecological animal husbandry in the Nordic countries.* Proceedings of NJF-seminar No. 303, Horsens, Denmark, 16-17th September 1999, Ed.s' Hermansen, J.E., Lund, V. and Thuen, E, pp45-50

Berry, E.C. (1994). Earthworms and other fauna in the soil. In: *Soil biology: Effects on soil quality.* Eds. J.L. Hatfield and B.A. Stewart, CRC Press, Boca, Raton, FL, pp61-90

Boorman, K.N. and Burgess, A.D. (1986). Responses to amino acids. In: *Nutrient requirements and nutritional research.* Eds. Fisher, C. and Boorman, K.N., Butterworths, London, 99-123

Clark, M.S. and Gage, S.H. (1996). The effects of domestic chickens and geese on insect pests and weeds in an agroecosystem. *American Journal of Alternative Agriculture* **11**:39-47

Clark, M.S. and Gage, S.H. (1997). The effects of free-range dometsic birds on the abundance of epigeic predators and earthworms. *Applied Soil Ecology* **5**:255-260

DEFRA-funded project OF0153. Effect of breed suitability, system design and management on the welfare and performance of traditional meat birds

Hughes, B.O. and Dunn, P. (1983). A comparison of laying stock: housed intensively in cages and outside on range. Research and Development Publication No. 18, West of Scotland Agricultural College, Auchincruive, Ayr

Kennel, W. (1990). The role of the earthworm *Lumbricus terrestris* in integrated fruit production. *Acta Horticulture* **285**:149-156

Larbier, M. and Leclercq, B. (1994). Translated by Wiseman, J. *Nutrition and feeding of poultry.* Nottingham University Press and INRA

Lubac, S. and Mirabito, L. (2001). Relationship between the activities of 'Red Label' type chickens in an outdoor run and external factors. *British Poultry Science* **42** (Supplement):S14

Maurer, V., Hirt, H. and Hordegen, P. (2000). Laying hen husbandry: effect of run management on turf quality. In: *Proceedings 13th International IFOAM Scientific Conference*, Eds. Alfoldi, T., Lockeretz, W. and Niggli, U., pp368

Mirabito, L., Joly, T. and Lubac, S. (2001). Impact of the presence of peach tree orchards in the outdoor hen runs on the occupation of space by 'Red Label' type chickens. *British Poultry Science* **42** (Supplement): S18

Mirabito, L. and Lubac, S. (2001). Descriptive study of outdoor run occupation by 'Red Label' type chickens. *British Poultry Science* **42** (Supplement): S19

National Research Council (NRC, 1994). *Nutrient requirements for poultry.* Ninth revised edition. National Academy Press, Washington

Poultry World (1959). The poultry handbook, 105

Raw, F. (1962). Studies of earthworm populations in orchards. I: Leaf burial in orchards. *Annals of Applied Biology* **50**:389-404

Robinson, L. (1948). *Modern poultry husbandry.* Crosby Lockwood, London

Teotia, J.S. and Miller, B.E. (1974). Nutritive content of house-fly pupae. *British Poultry Science* **15:**177-182

Thompson, A. (1952). *The complete poultryman.* Faber and Faber, London

UKROFS Standards (2001). Standards applying to livestock and livestock products from the following species: bovine (including Bubalus and Bison species), porcine, ovine, caprine, equidae, poultry. Chapter 1B

Yoshida, M. and Hoshii, H. (1978). Nutritional value of earthworms for poultry feed. *Poultry Science* **15:**308-310

8

ORGANIC PULLET REARING

Introduction

Probably the most contentious and technically challenging aspect of organic egg production is the rearing of replacement pullets.

There is increasing pressure within the UK and EC for organic egg producers to use organically reared pullets. Although EC legislation allows a transitional period whereby pullets up to 18 weeks of age may be brought in from conventional sources, this expires on 31 December 2003 (EC 1804/1999). In view of this MAFF provided funding for workshops during 2000 at ADAS Gleadthorpe, aiming to provide information on key technical problems associated with rearing pullets in an organic system. Attempts were made to provide possible solutions to these problems, and to highlight research priorities.

It soon became clear at these workshop discussions that there are likely to be far more technical difficulties in producing organic pullets than have been encountered in organic egg production. Some of the difficulties may be surmountable, but many are likely to increase costs. This chapter attempts to summarise the available scientific information, and, where possible, to quantify the costs of any necessary changes. Ideally the sensitivity of those

Based on MAFF funded workshops held in 2000, with contributions from Fiona Short[1], Claire Knott[2], Christopher Stopes[3], Arnold Elson[4] and Emma Garrett[4].
Advice and information on lighting for this chapter was received from Peter Lewis in 2001.

[1]ADAS Gleadthorpe
[2]Crowshall Veterinary Services
[3]EcoStopes Consultancy
[4]ADAS Consulting Ltd

costs to changes of technical input is required. However, it was deemed appropriate that cost should not be the first consideration, but rather that the wishes of consumers should be considered first, as is appropriate in any market driven sector. Consumers are concerned about variable standards for poultry and would prefer organic systems to be land based, probably in small flocks using mobile housing on well managed pasture, and integrated into a whole farm system (Stopes *et al.*, 2000).

Nevertheless costs must be estimated, since if the rearing of pullets to organic standards were to increase costs so much that the market for organic eggs disappears, then this would not serve the interests of either consumers or producers.

Photoperiodism as affecting maturity, productivity and welfare

Chapter 1 described the early history of the development of lighting programmes to provide year round supplies of eggs. The key principle to emerge is that it is the change in daylength rather than the absolute daylength which is influential. It had been recognised since the 1930s that the seasonality of egg production followed daylength.

For example, Whetham (1933), plotted daylengths and egg production for different latitudes, and concluded that the operative factor was the seasonal change in daylength and not the daylength *per se*. Other early authors such as Morris and Fox (1958a; 1958b) showed that it is important to avoid a declining daylength during lay, since this depresses ovulation. It was also recommended by these early authors that an increasing daylength during rearing should be avoided, because this brings about a precociously early sexual maturity, which results in smaller eggs, reduced yield, and the risk of prolapse, which, in turn, gives a higher risk of pecking, sometimes to the point of cannibalism. Morris published a series of comprehensive reviews (1968; 1994; 1999) of the work on this topic, including the physiological principles underpinning it.

In order to prevent problems associated with precociousness in winter- and spring-hatched non-organic pullets, the standard recommendation for non-light proofed houses in the UK, based on work in the above references, became a gradual step down in photoperiod from long photoperiods (e.g. 23 hours) at day old to reach the prevailing natural daily light period at

Table 1(a). SEASONAL NATURAL DAYLENGTHS IN BRITAIN AT LATITUDE 51°30'NORTH (LONDON)

Date	Daylength, hours and minutes	Increase on previous week, minutes	Date	Daylength, hours and minutes	Decrease on previous week, minutes
01 Jan	07:56	6	02 July	16:32	6
08	08:07	11	09	16:22	10
15	08:22	15	16	16:08	14
22	08:40	18	23	15:51	17
29	09:02	22	30	15:31	20
05 Feb	09:25	23	06 Aug	15:08	23
12	09:50	25	13	14:44	24
19	10:17	27	20	14:19	25
26	10:43	26	27	13:53	26
05 Mar	11:11	28	03 Sept	13:26	27
12	11:39	28	10	12:59	27
19	12:06	27	17	12:32	27
26	12:34	28	24	12:04	28
02 Apr	13:01	27	01 Oct	11:37	27
09	13:29	28	08	11:09	28
16	13:56	27	15	10:43	26
23	14:23	27	22	10:16	27
30	14:48	25	29	09:50	26
07 May	15:11	23	05 Nov	09:25	25
14	15:34	23	12	09:02	23
21	15:54	20	19	08:41	21
28	16:11	17	26	08:22	19
04 June	16:24	13	03 Dec	08:07	15
11	16:33	9	10	07:57	10
18	16:38	5	17	07.51	6
25	16:38	0	24	07:50	1

point-of-lay. The age at point-of-lay is usually recommended by the breeding company according to constraints such as body weight and frame size. Then, in the laying phase, the natural daily light period increased to a maximum of, e.g., 16 hours and 40 minutes in mid-late June at 51°30'N, (southern England). The artificial light period was increased so as to either match or slightly exceed the increasing natural light period. In July, the natural daily light period declines, but the artificial photoperiod was

Table 1(b) SUNRISE AND SUNSET TIMES (GMT) AT 51°30'NORTH

Date	Sunrise	Sunset	Date	Sunrise	Sunset
01 Jan	08:06	16:02	02 July	03:48	20:20
08	08:04	16:11	09	03:54	20:16
15	07:59	16:21	16	04:02	20:10
22	07:52	16:32	23	04:11	20:02
29	07:43	16:45	30	04:21	19:52
05 Feb	07:32	16:57	06 August	04:32	19:40
12	07:20	17:10	13	04:43	19:27
19	07:06	17:23	20	04:54	19:13
26	06:52	17:35	27	05:05	18:58
05 March	06:37	17:48	03 Sept	05:16	18:42
12	06:21	18:00	10	05:27	18:26
19	06:06	18:12	17	05:38	18:10
26	05:50	18:24	24	05:50	17:54
02 April	05:34	18:35	01 Oct	06:01	17:38
09	05:18	18:47	08	06:13	17:22
16	05:03	18:59	15	06:24	17:07
23	04:48	19:11	22	06:37	16:53
30	04:34	19:22	29	06:49	16:39
07 May	04:22	19:33	05 Nov	07:01	16:26
14	04:10	19:44	12	07:13	16:15
21	04:01	19:55	19	07:25	16:06
28	03:53	20:04	26	07:37	15:59
04 June	03:47	20:11	03 Dec	07:47	15:54
11	03:44	20:17	10	07:55	15:52
18	03:43	20:21	17	08:01	15:52
25	03:44	20:22	24	08:05	15:55

Source: Astronomical Applications Department, US Naval Observatory; *via* website of the Royal Observatory, Edinburgh. Quoted with permission.

For some responses the birds may not use exact sunrise and sunset times as cues to daylength. Indeed Yeates (1960) found some evidence that the interval between the start and end of civil twilight might be more relevant to behavioural responses.

held constant so that the hens did not experience a step-down in lighting during lay. An example, of such a step-down light programme for December-hatched non-organic pullets kept in southern England and receiving natural light is shown in Figure 1. (Note that in this and subsequent figures the daylength bars are based on the data in Table 1).

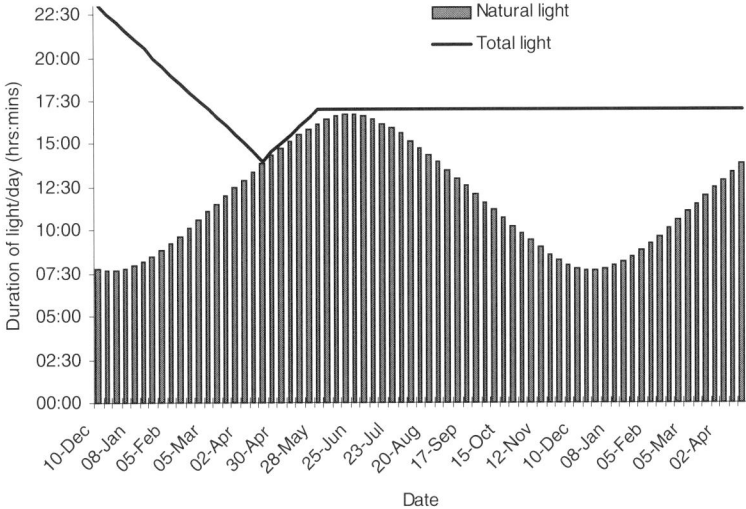

Figure 1. Example of a step-down lighting programme for application during rearing in December-hatched conventional pullets at about 51°N

There is a need to assess whether step-down light programmes may be applied to organic pullets. It has been assumed that there would be a requirement for organic pullets to be naturally ventilated, and that the natural light period may be supplemented using artificial lighting to provide a maximum of only 16 hours of light per day. This is according to (EC) 1804/ 1999 and UKROFS Standards. Additionally, irrespective of housing system, the proposed Welfare of Farmed Animals (England) Amendment Regulations states that the lighting regime shall include an adequate uninterrupted period of darkness lasting, by way of indication, about one third of the day. Thus, irrespective of the time-of-year that the chicks are to be hatched the maximum permissible daily light period at day old would be 16 hours.

For chicks hatched on 10 December and reaching point-of-lay at about 18 weeks of age the natural light period at point-of-lay would be about 13 hours and 55 minutes. In this example, it would be appropriate to reduce the light period from 16 hours down to 14 hours during brooding, and to then maintain the pullets on a 14-hour light period until point-of-lay (Figure 2). At point-of-lay the natural light period would be increased by about 25 to 30 minutes per week to a maximum of 16 hours 40 minutes in mid-late June. However to accord with the Directive, the artificial daily light period should be increased by 30 minutes per week until a maximum of only16 hours 0 minutes light per day is achieved. The artificial light period would then be maintained at 16 hours per day until the hens were depopulated.

However, the effect on ovulation rate of a further 40 minute step-up soon after the start of lay, due to increases in natural daylength, followed by a 40 minute step-down in natural daylength only a few weeks later, is not known, but it is likely to be minimal in most modern hybrids. Such a step down during lay contravenes earlier recommendations (e.g. Morris, 1968).

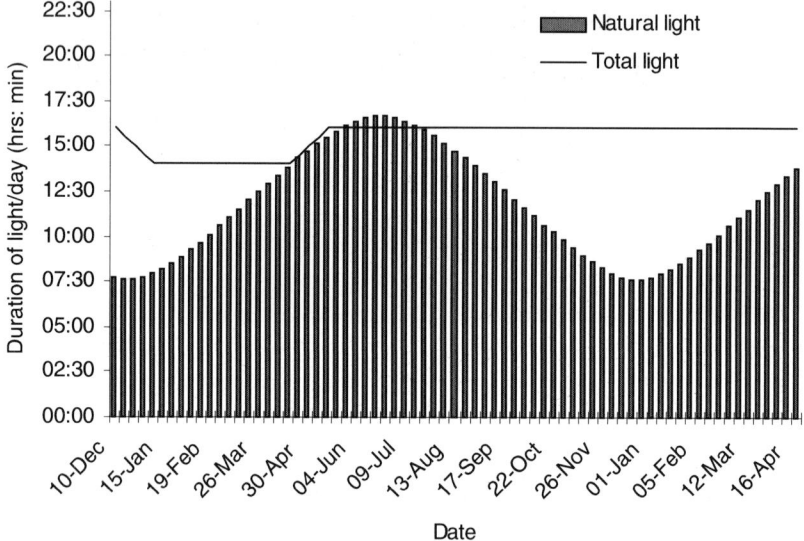

Figure 2. An example of a possible step-down light programme for applying during rearing in December-hatched organic pullets in southern England.

Perhaps a more problematic time-of-year for rearing organic pullets would be when chicks are hatched in early February and reach point-of-lay in mid-late June, when subsequently the natural light period is greater than 16 hours. For non-organic pullets reared in a non-light proof house it would be possible to apply a step-down light programme by reducing the light period from 23 hours at day old to 17 hours by 18 weeks. Then to stimulate sexual development, the light period may be increased by about 30 minutes each week until a daily light period of 19 hours is achieved. A 19-hour light period would then be maintained until the hens are depopulated (Figure 3). Alternatively, a 19-hour subjective daylength could be maintained by inserting a 6-hour period of light between hours 13 and 19.

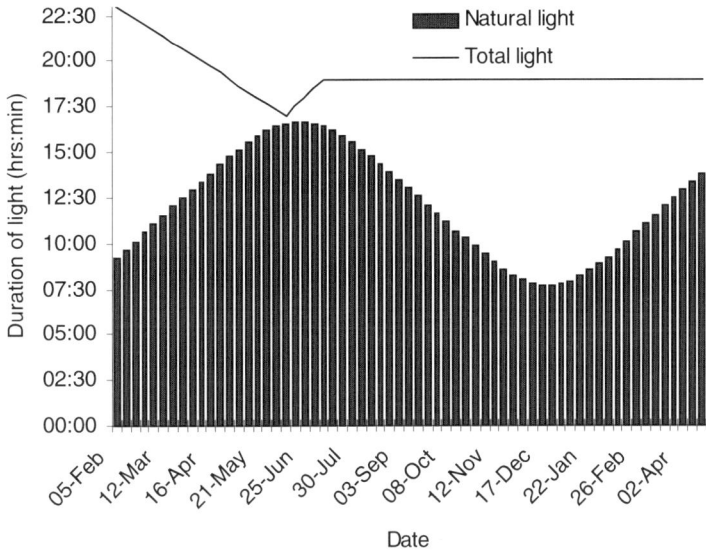

Figure 3. An example of a step-down light programme for application during rearing in February-hatched conventional pullets in southern England

It would not, however, be possible to photostimulate sexual development for February-hatched organic pullets, as light periods of greater than 16 hours are not permitted. Furthermore, it is not possible to step-down from 16 hours of light per day at day old because this duration of light would be needed to block the natural increase in light that occurs before the pullets reach point-of-lay. If the daily light period was reduced below 16 hours during brooding, the natural increment in daylength would stimulate precocious sexual maturity and this would increase the risk of prolapse. In February-hatched organic pullets it would only be possible to apply a 16-hour day during rearing and beyond (Figure 4), but little is known about the effects on egg performance of maintaining a constant long day from day old until the end of the laying year in modern hybrids. Furthermore, a 40-minute increase in natural daylength soon after the start of lay, followed by a 40-minute decrease in the natural daylength only a few weeks later, may override the constant 16-hour photoperiod and egg production may fall.

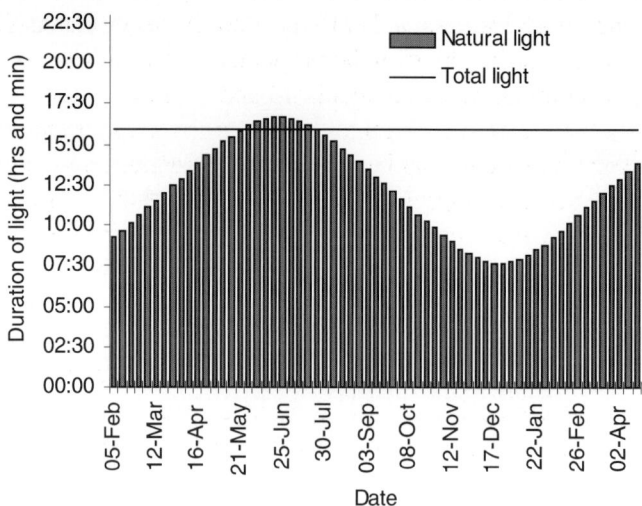

Figure 4. Example of the inability to apply a step-down light programme during rearing to February-hatched organic pullets in southern England

Whilst one apparently attractive approach is to try to light-proof the rearing house, so as to permit the control of the duration of light within the range of 8 to 16 hours per day, this is not practicable. It would be impossible to achieve adequate light exclusion (i.e. below the bird's published darkness threshold of 0.4 lux; see below) while yet providing sufficient ventilation for bird health, performance and welfare, and for preventing heat stress in warm weather. Anyway, organic standards require natural light to be available.

Morris and Owen (1966) published a photoperiodic threshold of 0.4 lux, above which the birds treated a period of dim light as day. This became the target degree of light proofing for many years. However, Emmans (1968) found that the application of light programmes had not completely eliminated effects of hatch date, and Spencer and Charles (1968) demonstrated that this was probably because there were substantial proportions of the houses of the time brighter than 0.4 lux when they were intended to be dark. Notwithstanding that Lewis *et al.* (1999) reported that supplementary light as low as 0.03 lux during the rearing phase would advance sexual maturity. Charles *et al.* (2002) concluded from a review that 0.4 lux was still a suitable practical recommendation.

In organic egg production, some producers may want to use traditional breeds rather than modern hybrids, especially as UK consumers are increasingly demanding novel niche products. However, it is possible that the use of traditional breeds in specialist versions of organic production would make the use of supplementary lighting more important rather than less. Practical experience, and experimental evidence from Gleadthorpe (Charles and Tucker, 1993) suggested that modern hybrids are so genetically predisposed to ovulate that they may be becoming refractory to lighting treatments. In contrast, Renema *et al.* (2001) observed that, when modern hybrids were reared at an intensity of 1 lux, age at sexual maturity was not significantly different from an old random-bred control, but ovarian and oviducal weights were significantly reduced. It was concluded that light intensity may be more, and not less, critical for modern hybrids. It is possible, therefore, that modern hybrids may respond differently to natural lighting than did earlier commercial hybrids. Lewis *et al.* (1997; 2000) found evidence of genetic differences in photosensitivity among different hybrids, whilst Lewis *et al.* (2000) observed an increased response, in terms of advance in age at first egg, to a change in daylength for succeeding generations of the same hybrid.

Morris (1968) published an equation, quoted below, to describe the effects of natural daylength on maturity. It permits some quantification of the sensitivity of age at first egg to daylength change for hybrids of that time.

$$y = \mu - 1.59(\Delta D)$$

where: y = age at first egg (days)
 μ = strain mean age at first egg (days)
 ΔD = total change in daylength occurring between hatching and the time of sexual maturity (hours).

Using the above equation, age at first egg for December-hatched non-organic pullets receiving a step-down in daylength from 23 hours to 14 hours during rearing (example illustrated in Figure 1) has been calculated to be 140 days, when using a strain mean for age at sexual maturity of 126 days.

For December-hatched organic pullets the maximum step-down in daylength is from 16 hours to 14 hours during rearing (example illustrated in Figure 2) and the age at first egg has been calculated to be 129 days. For February-hatched organic pullets there can be no step-down in daylength during rearing (example illustrated in Figure 4), and the age at first egg has been calculated to be 126 days.

Points that arise from these calculations are: 1) there will be fewer eggs but larger eggs when egg producers use December-hatched organic pullets, than when using December-hatched non-organic pullets, and: 2) the differences between egg production performance of non-organically reared pullets and pullets reared in an organic system are expected to be greatest in February-hatched flocks. This may lead to seasonality of demand for pullets reared to point-of-lay at other times of the year, and it may be a complicating factor when fully integrating organic pullet rearing or organic egg production into a whole farm system.

Costings of the effects of hatch date on egg production performance have not been calculated because the equation given by Morris (1968) requires a prediction of age at point-of-lay for calculating the change in daylength, and is therefore indeterminate.

For UK pullet rearers the inability to apply light programmes over a small part of the year, or the inability to markedly stimulate sexual maturity over a larger part of the year, is due to the restriction on daylength. If organic pullet rearers were able to use longer photoperiods than 16 hours then there would be more flexibility. This approach would also benefit organic pullet rearers in other Northern European countries. Organic pullet rearers in some Southern European countries may be less affected by the new Regulations as daylength during summer remains below 16 hours and even for February-hatched flocks a step-down in lighting during rearing would be possible (Figure 5).

Nutrition

PROTEIN SOURCES

For most of the conventional period the dominant protein sources have been based on soya beans from North America. However for reasons such as firstly aspirations for less dependence on imports, and secondly a

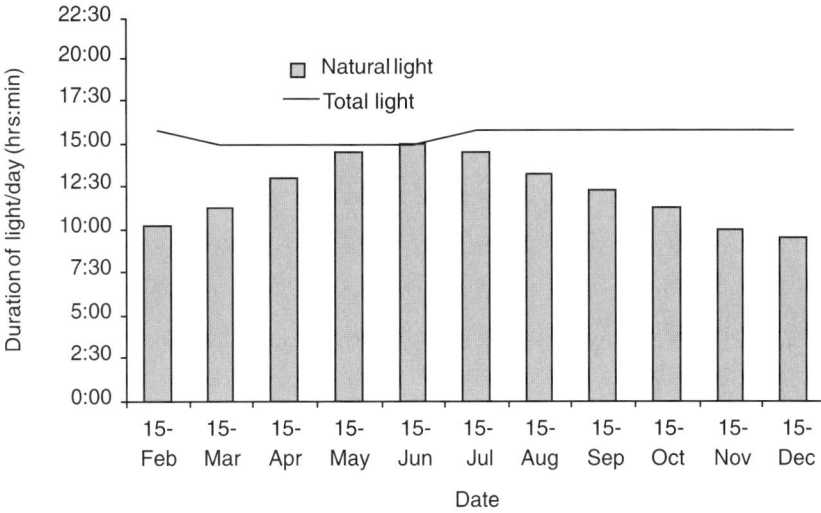

Figure 5. An example of a step-down lighting programme for February-hatched organic pullets reared at a latitude of 40°27'N (e.g. Madrid)

lack of consumer confidence in the segregation of non-GM soya, there has been interest in the possible use of more European home-grown proteins (see Chapter 3).

The biggest nutritional problem when formulating rations for organic pullets will be in meeting the birds' essential amino acid requirements for immunocompetence, maintenance and growth. Since 24 August 2000 it has not been permissible to include synthetic amino acids in organic poultry rations (EC 1804/1999). Synthetic methionine has been widely used in conventional poultry rations since the early 1960s and more recently synthetic lysine has also been used. All of the protein sources reviewed in Chapter 3 have lower lysine concentrations than full fat soya. Whilst rapeseed meal and sunflower meal have higher concentrations of methionine plus cystine than full fat soya, their inclusion rates are limited because of the presence and toxicity of ANFs, and because of low metabolisable energy value, respectively. As solvent extracted feed ingredients are not allowed in organic rations oil seed by-product meals would need to be produced by extrusion.

Oily fishmeal is allowed in organic starter rations and it has a high crude protein content and high essential amino acid contents relative to full fat soya. It is also a good supplier of calcium and available phosphorus. However, there are some consumers who want eggs from hens fed vegetable based diets, and so the acceptability of feeding fishmeal during early life may not be universal. An organic pullet feed was formulated without oily fishmeal and without synthetic lysine and methionine; using wheat, full fat soya, sunflower, potato protein, wheatfeed, plus vitamins and minerals. Although it was possible to achieve a dietary methionine content of 6.0 g/kg, comparable to that of a conventional chick ration, the crude protein and available lysine contents were much higher and the amino acid balance was poor.

PROTEIN QUANTITY AND QUALITY

The concept of an ideally balanced protein is sometimes used in estimating the requirements of the birds for essential amino acids (e.g. McDonald *et al.*, 1995). This concept is based on the proposition that the proportions of essential amino acids in an ideal dietary protein are the same as those in the metabolic body pool.

The supply of a perfectly balanced protein needed for the maintenance of growing broiler chickens has been calculated by Emmans and Fisher (1986); whose work was explained by Larbier and Leclercq (1994) as follows:

$8 \ (P_{max}^{-0.27} \times Pr)$ g/day per bird

where P_{max} = mature protein mass, kg
and Pr = body protein mass excluding feathers, kg, at the relevant age

Their calculation of the amount of an ideal protein needed for growth was $1.25 \times \delta$ prot g/d per bird; where δ prot = protein gain, g/d. Larbier and Leclercq (1994) suggested multiplying the results by 1.65 for practical feeds based on maize and soya.

The maintenance plus production modelling concepts developed for laying hens by Fisher *et al.* (1973) were applied to the growing chicken by Boorman and Burgess (1986). They applied 26 data sets for lysine, 13 for tryptophan, and 15 for the sulphur amino acids. The following expressions are based on some of their results for practical diets, (mg/g of gain and mg/day per bird respectively):

Lysine	14.86 δW + 82W
Tryptophan	2.70 δW + 6.8W
Sulphur amino acids	11.59 δW + 41W

Where δW = weight gain, g, and W = liveweight, kg.

The net efficiencies of amino acid utilisation for weight gain were given as 0.71, 0.79 and 0.60 for these three amino acids respectively. For comparison, in the expressions quoted above Emmans and Fisher (1986) had used 0.80 as the efficiency of utilisation of an ideal protein for growth.

Larbier and Leclercq (1994) published estimates of the amino acid content of chicken tissues and of requirement, based on the review paper by Boorman and Burgess (1986), (Tables 2 and 3 respectively).

Table 2. AMINO ACID CONTENT OF THE CARCASS AND FEATHERS OF GROWING CHICKENS (g/100g PROTEIN)

Amino acid	Carcass (g/100g protein)	Feathers (g/100g protein)
Lysine	6.5	1.6
Sulphur amino acids	4.0	7.9
Tryptophan	0.9	0.7
Threonine	4.2	4.6
Leucine	7.2	8.5
Isoleucine	4.3	6.4
Valine	4.7	8.9
Histidine	3.5	0.7
Arginine	6.8	7.3
Phenylalanine + tyrosine	7.0	7.4

Ideal protein profiles for broilers, expressed as fractions of the lysine allowance, were given by Peisker (1996). This approach permits subsequent research work to concentrate on lysine requirements (Mack *et al.*, 1999) (Table 4).

Table 3. ESTIMATES OF ESSENTIAL AMINO ACID REQUIREMENTS FOR GROWING CHICKENS

Amino acid	Maintenance (mg/d per kg liveweight)	Growth (g/100 g liveweight gain)
Lysine	82	1.49
Sulphur amino acids	60	1.16
Tryptophan	10	0.27
Threonine	86	0.75
Leucine	93	1.21
Isoleucine	58	0.77
Valine	70	0.95
Histidine	63	0.37
Arginine	50	1.40
Phenylalanine + tyrosine	370	1.20

Table 4. SOME AMINO ACID REQUIREMENTS OF BROILER CHICKENS AGED 20 TO 40 DAYS, AS PERCENTAGES OF THE LYSINE REQUIREMENT

Amino acid	% of lysine requirement
Methionine plus cystine	75
Threonine	63
Tryptophan	19
Arginine	112
Isoleucine	71
Valine	81

There is anecdotal evidence that a low methionine intake may be conducive to feather pecking. This seems possible, since Table 2 shows that feathers are rich in methionine.

Classen and Stevens (1995) reviewed some evidence for a reduction in the utilisation of the first limiting amino acid as the protein supply increased. However, Boorman (1992) suggested that while poor utilisation of a deficient amino acid in the presence of an excess of others was possible, poor response to low quality proteins was more likely to be through a reduction of feed intake. There have been estimates of the optimum concentration of lysine in the crude protein. Morris *et al.,* (1987) estimated it at 54 g/kg crude protein while the estimate of Rose and Salah Uddin (1997) was 66 g/

kg. The difference could have been due to changes in the body composition of broilers over the ten year interval between the two pieces of work.

The over supply of crude protein will not be a satisfactory way of coping with the withdrawal of synthetic methionine. The birds would need a higher water intake in order to excrete surplus nitrogen. This could lead to wetter droppings and consequent poor litter conditions.

ENERGY BALANCE

The energy balance of the chicken has been studied for a long time. For example Kleiber and Doherty (1934) were followed by later authors such as Freeman (1963), King and Farner (1964), Romijn and Lokhorst (1966), Richards (1974) and van Kampen (1981), all of whom studied chicken calorimetry. Pioneers of animal calorimetry and bioenergetics included Brody (1945) and Kleiber (1961), later appraised and reviewed by Blaxter (1989).

Brody (1945) found that across genotypes maintenance requirement was proportional to $A^{0.73}$, where A = mature weight. Using the work of such authors as a physiological basis Emmans and Fisher (1986) calculated that the energy requirements for the growth of chickens were 60.3 MJ per kg of protein retained.

Dietary energy requirement for growing leghorn pullets has been given by Leeson and Summers (1997) as 11.5 to 12.3 MJ/kg depending on age and dietary protein content.

It is clear from the calorimetric work that the metabolisable energy (ME) requirement of poultry is profoundly affected by environmental temperature, because of its effect on heat loss. The literature contains many references to the practical effects of this on feeding requirements for adult layers (e.g. Emmans and Charles, 1977), but there has been very little work on the practical temperature responses and requirements of growing replacement pullets. An exception was the work of Cowan and Michie (1983) who used high rearing temperatures in order to restrict growth. They compared 15, 20, 25 and 30°C. Body weight at 128 days of age was progressively reduced by increasing rearing temperature, and subsequent egg output was also progressively reduced.

NUTRITION AND IMMUNOCOMPETENCE

It is generally recognised that vitamin E, in synergy with the mineral selenium, has a part to play in immunocompetence in young birds (Larbier and Leclercq, 1994). It acts, among other things, as an antioxidant in pathways involving glutathione, and has a role in the synthesis of haem.

However vitamin E is not the only relevant nutrient. The n-3 fatty acids in fish oils have been associated with immunocompetence in several species (Prickett *et al.*, 1982). Fritsche *et al.* (1991) found that 70 g/kg of fish oil in the diet significantly enhanced antibody production in chicks. Korver and Klasing (1995) found that fish oil improved the resistance of broiler chicks to coccidia.

Tsiagbe *et al.* (1987) showed that high levels of methionine enhanced immune responses. They considered that extra methionine may be important for the synthesis of IgG antibodies, or perhaps as a T cell helper. The methionine requirement for health may be above that for growth. Unfortunately it is difficult to achieve satisfactory methionine concentrations in vegetable based diets.

Thus the concept of an interaction between nutrition and immunity is not a new idea, but a recent paper proposed a radical new aspect of the concept for very young chicks. Dibner (2000) argued that dependence on yolk sac reserves for the nutritional needs of the first few hours is insufficient. Yolk sac protein contains maternal antibodies; which are better used for immunity than as amino acid sources. She proposed the provision of feed within the chick transport boxes, for consumption before day old. For this purpose her company had produced a paste of very high protein content, sold under the trade name Oasis®.

This approach to very early nutrition is in stark contrast with some beliefs held during the historic period (before World War I). Hawk (1910) advised that chicks required no feed for the first 24 hours, and he claimed to have kept them without feed for two days without apparent ill effects. However, he may have been expressing a rather personal view, since he quoted the earlier advice of Sir Walter Gilbey, who recommended getting chicks off to a good start with the feeding of hard boiled eggs or bread and milk. This suggests that the need for a generous supply of a high quality protein at that stage of the bird's life had already been realised, though the reasons would not have been understood in terms of amino acid balance.

METHODS OF FEED PRESENTATION (INCLUDING CHOICE FEEDING)

During the conventional period poultry feeds, including the feeds offered to the birds during brooding and rearing, have nearly always taken the form of complete single ground and mixed feeds, though sometimes fed as pellets or crumbles. During the traditional period the dry feed was often mixed with a little water to form a wet mash. The term mash was, and sometimes still is, used to include dry feeds. Hawk (1910) considered that dry feeding could cause contracted crops.

In the traditional period it was usual to offer laying hens separate feeding of a calcium source, and Howes (1939) went so far as to insist that grit and shell should always be available, but that they should not be mixed with the feed. Formal experiments on choice feeding go back a long time, and Graham (1934) offered caged layers a choice between whole maize, whole oats and a mash, as well as the separate provision of oyster shell. Individual birds laying at high rates chose a higher protein mixture. Dove (1935) found that chicks chose an almost perfectly balanced diet when offered a multiple choice.

During the conventional era serious interest in choice feeding probably dates from the work of Emmans (1977), who found that layers were able to make appropriate choices between a whole feed and ground barley. Several authors found evidence that layers could select for calcium, as well as energy and protein, (e.g. Mongin and Saveur, 1974), but Emmans (1974) and Leeson and Summers (1978) recognised that there may be interference between calcium and protein appetite drives, so that at least three choices may be desirable, which is interestingly redolent of the work of Graham (1934) and Dove (1935) in the previous era. A great deal of work on choice feeding in various classes of poultry has continued since the 1970s, (e.g. Cowan and Michie, 1978; Sinurat and Balnave, 1986 and Gous and Swatson, 2000), usually with encouraging results. MacLeod and Dabutha (1997), working with quail, added the important observation that choice feeding helped to maintain growth rate over a wide range of temperatures, by allowing the birds to maintain protein intake. Forbes and Covasa (1995) considered the use of whole cereals in detail. The use of whole cereals can potentially save milling costs, but because of the need for flint grit the conventional industry has feared problems with abrasion damage to machinery in the slaughter plant. There may also be microbiological contamination hazards (see below).

In the conventional era feed trough design difficulties have usually prevented the successful commercial application of choice feeding. This may be an example of an aspect of poultry technology in which beneficial techniques from the traditional era have been lost.

NUTRIENT INTAKE AND GROWTH

One of the reasons for the experimental success of choice feeding, noted by several of the above authors, has been its ability to allow the bird to optimise nutrient intake despite changes in appetite of *ad libitum* fed birds due to factors such as temperature. Practical difficulties with the optimisation of nutrient intake have a long history of their own, which may be relevant to the question of nutrition and the feasibility of producing well grown pullets. This was of concern to one of the workshops on rearing organic pullets.

The formulation of whole mixed diets by matching the analysis of a mixture of available ingredients to published estimates of requirement has been standard feeding practice for the whole of the conventional period. Computers have been used to achieve this since the early 1960s, and least cost diet formulation was thus one of the earliest examples of the application of computers to industrial problems. However, the procedure works on optimising diet *composition*, while birds respond to nutrient *intake*. Therefore unless the formulator has knowledge of the voluntary feed intake of the particular flock being fed, nutrient intakes may not necessarily be correct.

Morris (1986) commented that this is a concept which has been rediscovered about every ten years since Combs (1962) showed that dietary energy value affected the requirement for feed protein content, thus introducing the idea of an energy:protein ratio. Payne (1966) made the important point that there is a need to adjust protein concentrations for *ad libitum* fed layers following a change in environmental temperature, as well as for different dietary energy levels. This is because temperature affects energy requirement, and thus voluntary feed intake, while protein, vitamin and mineral requirements are, for practical purposes, unchanged by temperature. Morris called this important interaction between environment and nutrition *Payne's hypothesis*. It could be relevant to changes in the husbandry of pullets and we may need to discover it yet again. Charles and Payne (1966b) found that factors other than dietary energy and temperature could influence the feed formulation requirements through *ad libitum* feed intake, and suggested the term "environmentally induced nutritional stress".

Kwakkel (1999) pointed out that a target body weight is not a sufficiently detailed criterion for describing the adequacy of physiological development required during rearing in order to sustain egg production. He postulated, and found some evidence, that each body organ and tissue followed its own maturation curve. This, he suggested, resulted in a variation in nutritional demand over time. The reproductive tract was observed to show a growth spurt late in rearing. A particular amount of body protein (about 187 g/bird in his birds) seemed to be associated with the onset of sexual development. His worked confirmed the classical growth priority order of bone, muscle, fat as described by Hammond (1940) and others for all farm animals. Changing nutritional demand over time could be construed as a case for choice feeding, since without sample slaughter it would be difficult to know which parts of the body were experiencing growth spurts. Govaerts *et al.,* (2000) confirmed the notion of changing growth priorities. They found that during feed restriction early in the life of broilers, i.e. at four days of age for four or eight days duration, the birds gave priority to the development of the supply organs such as the stomach, at the expense of muscle.

COMPETITIVE EXCLUSION AND PROBIOTICS

The healthy chicken gut is far from sterile. It could be regarded as a habitat. Ewing and Cole (1994), in a review book, listed twelve bacterial genera routinely found in the chicken gut. The numbers of individual organisms are impressive, and in the gut of the healthy pig Ewing and Cole tabulated approximations of 10^6 to 10^9 *Lactobacilli* per g of fresh weight of gut contents, and 10^4 to 10^8 coliforms.

A healthy growth of the non-pathogenic flora helps to prevent colonisation by disease forming types. This is sometimes called competitive exclusion because there are only a finite number of colonisation sites available (Ewing and Cole, 1994). *Lactobacilli* were considered to compete with *E. coli* and *Salmonella*. The notion that intestinal bacteria play a part in health is old and dates from Metchnikoff (1908). The term competitive exclusion was first used by Greenberg (1969), and Nurmi and Rantala (1973) first applied the concept to poultry. In their fascinating work on chicks in Finland they found that giving chicks adult gut contents inhibited the establishment of *Salmonella infantis*, which was a troublesome *Salmonella* species in the Finnish broiler industry at the time. They blamed the *Salmonella* colonisation on the abnormally hygienic growing conditions hampering the development of the intestinal flora. Their hypothesis was that newly hatched chicks are in a transitional phase between germ-free and normal animals for a few days, during which time their defences are weak.

Lactobacilli have often been the basis of probiotics and Jin *et al.*, (1997) listed the modes of action as including antagonism to pathogens (through the action of bactericidal organic acids and hydrogen peroxide), competitive exclusion, aggregation with pathogens, suppression of ammonia production, neutralisation of enterotoxins produced by pathogens and possibly immuno-stimulation. Edens *et al.*, (1997) demonstrated that the administration *in ovo* of *Lactobacillus reuteri* increased its rate of intestinal colonisation and decreased the colonisation by *Salmonella* and *E. coli* in chicks and poults.

A related phenomenon is competitive binding. Ewing and Cole (1994) have described the use of oligosaccharides such as mannose to bind organisms by exploiting the adhesion of the lectins of the organisms to carbohydrates. For example Ofek *et al.*, (1977) found that D-mannose and methyl α-D-mannopyranoside inhibited adherence to epithelial cells by two strains of *E. coli* but not by *Streptococci*. Later Oyofo *et al.*, (1989) found that 2.5% mannose in the drinking water of broilers significantly reduced the level of colonisation of *Salmonella typhimurium* in the caeca.

Charles and Bray (1998) reviewed such topics in the context of problems of turkey enteritis, and some of the above is taken from their review.

Housing facilities and associated services

The design and management of housing for poultry need both to fulfil the requirements of organic standards and to allow for an efficient, but nevertheless welfare orientated, management of animals (Lampkin, 1997). A system that may be suitable for the general farm or small holding, where organic poultry production is merely a part of the farmer's activities, may be unsuitable for the larger business (Robinson, 1961).

Under organic standards it is intended to provide the birds with an environment in which all normal behaviours can occur. This will minimise the amount of stress and hence be advantageous to health and productivity. Outdoor poultry are shut in at night, in order to protect them from predators, and therefore the design of the building must take into account the behavioural needs of the birds during this period of confinement.

To conform with organic standards natural lighting must be available. However, the presence of such light is believed to increase feather pecking and cannibalism. In designing the windows of the houses therefore, consideration must be given to ensuring that uniform distribution of light can be achieved.

The act of finding food is an important social activity for hens. All birds in a flock normally feed at the same time, so that the sounds of pecking and scratching act as stimulants for other birds. Similarly feeding behaviour can be stimulated by the sound of mechanical feeders running. Appropriate space must be provided for the birds within the housing system in order that such activities can be carried out. Feed and water are provided in the house, although some egg farmers provide a scratch feed of grain on the range to encourage the birds to go out and to move them to different parts of the range (Elson, 1995).

Furthermore, housing needs to provide sufficient space for flying and fluttering and to incorporate a lighting system that allows adequate resting periods. Bathing in sand and dust not only increases the comfort of the birds, but also maintains hygiene and helps to reduce the number of ectoparasites. Areas to allow such activity are needed either within the housing system or in the range area. Outside areas also need to include shelter from predators, such as birds of prey.

There are two main approaches to housing used by organic producers: mobile units which can be moved to utilise the pasture in rotation, and static housing, whereby the birds are allowed access to an outside area covered by vegetation (Lampkin, 1997).

MOBILE HOUSING

There are now a number of suppliers offering mobile housing units suitable for organic production. Depending on the type of mobile housing system purchased, equipment such as feeders and drinkers may be included. In other instances these may need to be purchased as extras. There are no common standards for the provision of equipment. However, it is generally recognised that an even distribution of feeders and drinkers is as important as adequate provision (Elson, 1996).

One advantage of mobile housing is that the birds can be moved to fresh grass periodically, thereby reducing the risk of soil-borne parasites in the outside area. However, stock may still be re-infected by pathogens from their own droppings if these are retained in the house.

Although not needing planning permission in UK at present, one of the constraints on these units is their size in order to remain mobile. Units available commercially vary in capacity from 200 to more than 1000 birds. A further disadvantage is the need for all other production inputs (feed, straw, water and electricity) to be transported to and from the moveable units. This increases the need for, and therefore the cost of, labour. Overall, the cost per bird of using mobile housing systems is higher than the corresponding costs in static houses.

The design of these houses, and the materials used in their construction, vary. One common important factor is the need for insulation to maintain feed conversion rates in the winter. At cleaning out it is necessary to rest the house for about 7 days, to adopt a thorough cleaning programme, and to use an approved disinfectant, or preferably steam cleaning, depending on the construction of the house.

STATIC HOUSING

An advantage of larger static housing systems is the opportunity to introduce mechanisation (or partial mechanisation) for the provision of feed, water and collection of droppings. This leads to a lower cost per bird than in mobile systems. Controlled environment housing is not possible because of the provision of pop holes.

Management of the outside area tends to be more of an issue with static houses (Lampkin, 1997). An element of rotational grazing needs to be adopted in order to reduce the risk of parasites and diseases in the soil, and to maintain the vegetation cover. The area immediately adjacent to the house is utilised to the greatest extent. Hence, there is a need for management practices to be adopted (such as putting straw down in this area) to reduce the amount of mud that is carried into the sheds in wet weather, since the mud may contaminate fittings. Housing is best configured so that the birds enter the pop holes over a perforated floor area, in order to clean their feet before entering the house (Elson, 1995).

Static housing, as currently used in organic production, varies considerably and so do the flock sizes housed within the units. Reducing the flock size *per se* will increase capital costs, due to proportionally higher building and equipment costs.

On a conventional farm it is unlikely that the existing houses will be suitable for organic production. Ex-intensive units will tend to have been built for housing large numbers of birds: an option not permitted under the organic standards. Furthermore, the proximity of the houses to each other acts as a restriction on the amount of land available to be utilised as the range area.

For producers wishing to build new Class 3 buildings for organic production, planning permission will be needed (unless the gross floor area is less than 465 m^2). Further criteria of distance from roads, other poultry units and dwelling houses will also need to be met.

Light intensity, feathering and bird behaviour

LIGHT INTENSITY

The management of light is an important issue for all types of poultry production, whether intensive, or extensive. With extensive systems such as organic, it is difficult to control the intensity. Intensities of about 10 lux have usually recommended for egg production and are still recommended (Charles *et al.*, 2002). In intensive systems intensities much over 10 lux are usually avoided in order to decrease feather pecking.

Some have suggested that natural light is necessary for laying hens (Huber and Fölsch, 1985a; 1985b). There is no evidence to support this although some aspects of natural light variation can help the birds to adapt to their environment (Elson, 1998).

However, bright light can also have adverse effects on the behaviour of the birds. It increases activity, and hence tends to worsen feed conversion efficiency. Furthermore, and of growing concern, is the effect of light intensity on aggression and feather pecking, and the risk of cannibalism (Hughes and Duncan, 1972). This has become more important with increased interest in bird welfare.

FEATHER PECKING

The significance of light quality in relation to the welfare of poultry is still obscure and debatable.

Generally, feather pecking is found to start during the rearing phase (Huber-Eicher, 2000). A study of feather pecking in Switzerland found that over a third of farms surveyed had pecking problems during rearing. The consequential losses occurred in the laying period. The problems suffered were attributed to a lack of care and an increase in stocking density. A stocking density of more than 10 birds/m^2 was found to increase feather pecking, and crowding occurred around the feeders and drinkers.

It is normal behaviour for birds to peck others gently whilst they are dustbathing (Keeling, 2000). Day old chicks peck in the search for food (Spicer, 2000).

Feather pecking alone may not necessarily be a problem, despite leading to a partial loss in feathers and resulting in poor bird appearance. The real problem arises when it leads to cannibalism, where birds inflict injuries, mainly to the cloacal region. This leads to health problems due to infection, and it also has a negative effect on productivity. Poor feather is also associated with poorer cold tolerance (see Chapter 4).

Feather pecking can be caused by a variety of factors, such as exposure to sunlight, insufficient protein in the diet, excessive egg size, large flock size, stress and boredom (Ekstrand, 1996; Fölsch *et al.*, 1992, Bauer and Keppler, 1996). A recent workshop held at the University of Bristol School of Veterinary Science explored solutions to the problem of feather pecking in laying hens. The message from the majority of papers was the necessity to avoid high stress levels. Not only is stress the main factor that increases feather pecking but also it decreases immunocompetence (Huber-Eicher, 2000).

Pasture management and hygiene

SOME HYGIENE PRINCIPLES

Knott (2000) pointed out that viruses may be spread by contact or by aerosol, including windborne spread, depending on the type of virus. Some are fragile outside the host, but others can persist for long periods. They can also be introduced onto a site by visitors, rodents, wild birds, pets, insects and equipment. Bacterial infection may be spread similarly, but it is less likely to be windborne, particularly over long distances. Contaminated feed and water can introduce bacterial contamination. Some bacteria, such as *Mycoplasma,* survive only a short time outside the host, but others survive for long periods, particularly if protected by organic matter.

Unfortunately, outdoor systems of poultry production are inevitably more vulnerable to some sources of hazards from disease and food poisoning organisms. This is because they are more exposed to contamination by sources such as wild birds and rodents. Also the chickens are able to contact their own droppings. One of the objectives behind most designs of indoor controlled environment designs during the conventional period was the exclusion of wild birds and rodents. An original objective of battery cages, proposed during the traditional period, (after their introduction into

Britain about 1925), was the prevention of the contamination of feed and water by droppings (Robinson, 1948).

Permin and Nansen (1996) reviewed some diseases encountered in organic poultry systems in Denmark, and noted that they included some which had not recently been major problems in intensive systems. They included *Pasteurella multocida* infection, adenovirus infection and Newcastle disease, as well as various nematodes and endo- and ecto-parasites. *E. coli* secondary infection following cannibalism had occurred. By contrast, broilers in Sweden fed diets not containing coccidiostats had higher oocyst counts in litter and faeces than in ionophore treated birds, but untreated birds did not have clinical coccidiosis (Ekstrand *et al.,* 1994).

In an abattoir survey in Ontario, Canada, Herenda and Jakel (1994) found a low rate of pathological lesions in free range chickens. Georgot (1995) surveyed non-industrial production experience in France, including the principle diseases.

MANAGEMENT PRACTICES

History suggests that the control of parasites and enteric diseases has always been a priority aspect of good husbandry on range. The textbooks of the traditional period stressed this (e.g. Robinson, 1948). More recently, studies of the effects of factors such as distance from the house and the degree of faecal contamination of the land on parasitic status have been published by Bray and Lancaster (1992). Hane *et al.*, (2000) quantified the magnitude of the parasite challenges in modern free range egg production in Switzerland. Worm eggs and coccidial oocysts are resistant to normal disinfectants, and in the right conditions can be very persistent in the soil or on the floors of buildings (Knott, 2000). For example round worm eggs may survive in the soil for over a year. Adequate rotation is clearly an essential aspect of pasture management.

Pasture management and sustainability

Pasture management for pullet rearing involves the appropriate utilisation of pullet manure within the rotations on the whole farm, but without run off and without exceeding the manure nitrogen application limits imposed by Directive 91/676/EEC. The quantities of manure were given by Smith *et al.,* (2000), but the utilisation of manure on the land is discussed in more detail in Chapters 10,11 and 12.

Water supply

Arrangements can be made for the supply of water to outdoor units, but chickens on range will drink from puddles, raindrops on branches, small ponds and similar potentially contaminated sources (Elson, 2000; Knott, 2000). Standing water and puddles are attractive to birds on range and may result in some birds ingesting high levels of potentially pathogenic bacteria and parasites, though the risk is probably small (Knott, 2000).

Food safety

There can be little doubt that one of the key concerns that modern food consumers have is food safety. This is reflected in a great deal of concern in the trade and in the food chain. At a strategic poultry industry event in 1997 food safety was one of the three most important recurring themes stressed by food industry speakers (BOCM PAULS Ltd, 1997).

Fortunately the microbiological and practical aspects of the safety of poultry products have been well documented, and the following highlights are taken from a recent authoritative review source (Mead, 1998).

He considered that the most important food safety issues for the poultry industry were related to contamination by *Salmonella* and *Campylobacter* species. *Eschericia coli* 0157 had not been found on poultry meat despite the ability of the organism to colonise the alimentary tract of young chicks in experimental studies. Application of the Hazard Analysis Critical Control Point (HACCP) principle throughout the food chain was considered the most systematic and effective approach to food safety practice.

He noted that the rigorous biosecurity measures which can be imposed in controlled environment cannot be imposed on free range, though once *Salmonella* and *Campylobacter* gain entry spread can be rapid under close confinement systems. It was considered important to minimise visitors and to keep out other livestock and wild animals with a perimeter fence. Untreated feeds, such as whole wheat supplements, were regarded as hazardous compared with feeds which receive a heat treatment at the mill. Care was required with the cooling of pelleted feeds if the cooling processes involved drawing in air. As mentioned above contaminated drinking water constitutes a hazard for outdoor systems.

Discussion

There are currently no standards for organic pullet rearing, but presumably in due course standards will be proposed akin to those for organic table birds and in part to those for organic laying hens. Reference to organic table bird production standards would seem reasonable, as table birds, like replacement pullets, are sexually immature and growing. However, physical and nutritional requirements may differ between breed types and there is the additional need to consider the reproductive development of the pullet. Table birds are slaughtered prior to the onset of sexual maturity, and slow growth to this age is desirable so that the birds are not too large for consumer preferences and so that meat flavour is more developed. By comparison, egg type breeds are slow growing and they do not need low specification diets to retard growth; rather they need nutrient intakes that are suited to optimising growth, the development of the immune system and the development of the reproductive tract.

High dietary cereal contents would dilute the crude protein and essential amino acid contents and this may not allow optimal bird development. In the absence of synthetic amino acids there is an increased need to select feed ingredients other than cereals on a nutritional merit basis, as well as on a cost basis, but this is likely to be difficult as the range of ingredients is limited.

Pullet rearing may be considered in two parts: namely the brooding period and the growing period. Brooding may be carried out by specialist brooders, followed by a move to specialist grower sites, or the two processes may be carried out together on one site. The point of lay pullets may then enter egg production facilities on the same site, or they may be sold to egg producers.

Specialist brooder facilities may offer a better start for the chicks, as controlled environment facilities are able to provide a stable, optimal thermal environment despite changes in outside conditions. The houses would be static and because of an electricity supply they could be steam cleaned and this would help to minimise early disease challenges. A verandah attached to the brooder house may be used to acclimatise young birds to outside conditions. The growing facilities may be either static or mobile. Perches would need to be provided so that the birds have early experiences of perching. It would be important to avoid having the pullets in the growing facilities at the start of lay as nest boxes would not be available and this would encourage floor laying.

Even if specialist brooding facilities are not used it seems likely that chicks destined for organic egg production will be transferred at least once, namely at the start of lay when pullets are moved to laying facilities. The concept of brooding chicks in a given facility and then the pullets remaining in that facility until the end of lay does not seem realistic. It would involve the houses being fitted with brooding equipment, perches and nest boxes, as well as feeders and drinkers, and in small scale houses the equipment would soon clutter the available space. Litter would have to be replaced so as to avoid the excessive accumulation of droppings, the build up of noxious gases and bacterial loads. The continuity of egg supply and egg income would also be affected. Furthermore, in a whole farm system there would be considerable demands on the producer for husbandry and technical skills across the entire repertoire of brooding, growing and laying.

It is generally accepted that moving poultry is undesirable as it causes fear and physiological stress, but realistically this seems to be the only practical option. Strategies should be in place to minimise fear, to minimise handling damage and to safeguard the birds' thermal comfort during transport.

There would be welfare benefits associated with the development of specialist brooder, grower and egg production enterprises, as chick mortality is likely to be lower, and the birds' needs during the different stages of production are likely to be better understood and therefore better met. There may also be opportunities to co-ordinate specialist growers and specialist egg layers with organic arable producers, so that when using mobile housing facilities these enterprises can be integrated with arable rotations such that there is additional income on the farm, yet without allowing the land to become fowl sick (see Chapters 10, 11 and 12).

References

ADAS (1979)(first published much earlier, date unknown, probably 1940s) *Domestic poultry keeping.*

Bauer, M. and Keppler, C. (1996) Federpicken - die Krux der Legehennenhaltung. *Bioland* 1996 (5): 15-17

Beaumont, J. (1997) *The consumer has spoken.* Keynote address. BOCM PAULS Conference, The Belfry

Blaxter, K.L. (1989) *Energy metabolism in animals and man.* Cambridge University Press, Cambridge, UK

Boorman, K.N. (1992) Protein quality and amino acid utilisation in poultry. In: *Recent advances in animal nutrition.* Edited by Garnsworthy, P.C., Haresign, W. and Cole, D.J.A., Butterworth Heinemann, Oxford, UK, pp. 51-71

Boorman, K.N. and Burgess, A.D. (1986) Responses to amino acids. In: *Nutrient requirements and nutritional research.* Edited by Fisher, C. and Boorman, K.N., Butterworths, London, UK, pp. 99-123

Bray, T.S. and Lancaster, M.B. (1992) The parasitic status of land used by free range hens. *British Poultry Science* **33:** 1119-1124

Brody, S. (1945) *Bioenergetics and growth.* Reinhold, New York, USA

Charles, D.R. (1986) Temperature for broilers. *World's Poultry Science Journal* **42:** 249-258

Charles, D.R. (1996a) A brief history of the UK poultry industry. In: *UK Branch 50th Anniversary.* Edited by. Fisher, C. and Hann, C.M., World's Poultry Science Association, pp. 45-51

Charles, D.R. (1996b) The Museum of the British poultry industry. *Journal of the Royal Agricultural Society of England* **157:** 197

Charles, D.R. and Bray, T.S. (1998) Turkey enteritis and wet droppings - a review of the biological background. *ADAS Broiler and turkey conference.* Chilford

Charles, D.R., Lewis, P.D. and Tucker, S.A. (2002) Lighting programmes for laying hens. In: *Poultry environment problems: a guide to solutions.* Edited by Charles, D.R. and Walker, A.W., Nottingham University Press, Nottingham, UK

Charles, D.R. and Payne, C.G. (1964) The effects of ammonia on the performance of laying hens. *World's Poultry Science Association 2nd European Poultry Congress,* Bologna, pp. 109-112

Charles, D.R. and Payne, C.G. (1966a) The influence of graded levels of atmospheric ammonia on chickens. 1. Effects on respiration and on the performance of broilers and replacement growing stock. *British Poultry Science* **7**: 177-187

Charles, D.R. and Payne, C.G. (1966b) The influence of graded levels of atmospheric ammonia on chickens. 2. Effects on the performance of laying hens. *British Poultry Science* **7**: 189-198

Charles, D.R. and Tucker, S.A. (1993) Responses of modern hybrid laying stocks to changes in photoperiod. *British Poultry Science* **34**: 241-254

Charles, D.R. and Tucker, S.A. (1997) The poultry industry in the United Kingdom. *Journal of the Royal Agricultural Society of England* **158**: 175-183

Classen, H.L. and Stevens, J.P (1995) Nutrition and growth. In: *Poultry production*. World Animal Science sub-series C9. Edited by Hunton, P., Elsevier, Oxford, UK, pp. 79-99

Coles, R. (1966) Size changes in laying flocks. *The poultry review*. May and Baker Ltd., Dagenham, UK

Combs, G.F. (1962) The interrelationship of dietary energy and protein in poultry nutrition. In: *Nutrition of pigs and poultry*. Edited by Morgan J.T. and Lewis, D., Butterworths, London, UK, pp. 127-147

Cotterill, O.J. and Winter, A.R. (1953) Some nitrogen studies of built up floor litter. *Poultry Science* **32**: 365

Cowan, P.J. and Michie, W. (1978) Environmental temperature and choice feeding of the broiler. *British Journal of Nutrition* **40**: 311-315

Cowan, P.J. and Michie, W. (1983) Raised environmental temperature and food rationing as means of restricting growth of the replacement pullet. *Poultry Science* **24**: 11-19

Dibner, J. (2000) Early nutrition in young poultry. In: *Recent advances in animal nutrition*. Edited by Garnsworthy, P.C. and Wiseman, J., Nottingham University Press, Nottingham, UK, pp. 73-88

Dove, W.F. (1935) A study of the individuality in the nutritive instincts and the causes and effects of variation in the selection of food. *American Naturalist* **69**: 469-544 (Abstract)

EC 1804/1999 (1999). Council Regulation supplementing Regulation (EEC) No 2092/91 on organic production of agricultural products and indications referring thereto on agricultural products and foodstuffs to include livestock production.

Edens, F.W., Qureshi, R.A., Parkhurst, C.R., Qureshi, M.A., Havenstein, G.B. and Casas, I.A. (1997) Characterisation of two *Eschericia coli* isolates associated with poult enteritis and mortality syndrome. *Poultry Science* **76**: 1665-1673

Ekstrand, C. (1996) Inhynsningssystem : *Månge alternativ. Forskningsnytt om Æekologisk landbruk i Norden.* 1996 **4**, 9-11

Ekstrand, C., Algers, B., Thebo, P. and Hooshmand, R.P. (1994) Rearing broilers without coccidiostats. *Proceedings 8th International Congress on Animal Hygiene,* St. Paul, Minnesota (Abstract)

Elson, H.A. (1995) Environmental factors and reproduction. *Poultry production.* World Animal Science. Edited by Hunton, P., Elsevier, pp. 389-409

Elson, H.A. (1996) Egg production systems in the EU. Animal health and product quality. DLG-Verlag, Frankfurt, 12-14

Elson, H.A. (1996) *The poultry production guide.* Elsevier

Elson, H.A. (2000) Unpublished observations

Emmans (1968) Effective environmental control. 1. Evidence of the failure to control daylength under practical conditions. *World's Poultry Science Journal* **24:** 317

Emmans, G.C. (1977) The nutrient intake of laying hens given a choice of diets, in relation to their production requirements. *British Poultry Science* **18:** 227-236

Emmans, G.C. and Charles, D.R. (1977) Climatic environment and poultry feeding in practice. In: *Nutrition and the climatic environment.* Edited by Haresign, W., Swan H., and Lewis, D., Butterworths, London, UK, pp. 31-49

Emmans, G.C. and Fisher, C. (1986) Problems in nutritional theory. In: *Nutrient requirements of poultry and nutritional research.* Edited by Fisher, C. and Boorman, K.N., Butterworths, London, UK, pp. 9-40

Ewing, W.N. and Cole, D.J.A. (1994) *The living gut.* Context, Packington, UK

Fairbanks, F.L. and Rice, J.E. (1924) Artificial illumination of poultry houses for winter egg production. *Cornell Extension Bulletin* 90 (Abstract)

Farm Animal Welfare Council (1997) *Report on the welfare of laying hens*

Fisher, C., Morris, T.R. and Jennings, R.C. (1973) A model for the description and prediction of the response of laying hens to amino acid intake. *British Poultry Science* **14:** 469-484

Fölsch, D.W. and Hoffmann, R. (1992) Beratung Artgerechte Tierhaltung (eds.) Artgemässe Hühnerhaltung (Appropriate poultry husbandry) *Alternative Konzepte,* 79. Müller, Karlsruhe, Germany

Forbes, J.M. and Covasa, M. (1995) Application of diet selection by poultry with particular reference to whole cereals. *World's Poultry Science Journal* **51**: 149-166

Freeman, B.M. (1963) Gaseous metabolism of the domestic chicken. 4. The effect of temperature on the resting metabolism of the fowl during the first month of life. *British Poultry Science* **4**: 275-278

Fritsche, K.L., Cassity, N.A. and Huang, S-C. (1991) Effect of dietary fat source on antibody production and lymphocyte proliferation in chickens. *Poultry Science* **70**: 611-617

Georgeot, P. (1995) The biological chicken. *Agricoltura-Biologica Supplement 2*, 1-3 (Abstract)

Gous, R. and Swatson, H.K. (2000) Mixture experiments: a severe test of the ability of a broiler chicken to make the right choice. *British Poultry Science* **41**: 136-140

Govaerts, T., Room, G., Buyse, J., Lippens, M., de Groote, G. and Decuypere, E. (2000) Early and temporary quantitative food restriction of broiler chickens. 2. Effects on allometric growth and growth hormone secretion. *British Poultry Science* **41**: 355-362

Graham, J.C. (1934) Individuality of pullets in balancing the ration. *Poultry Science* **11**: 34-39

Greenberg, B. (1969) *Salmonella* suppression by known populations of bacteria in flies. *Journal of Bacteriology* **99**: 629-635

Hammond, J. (1960) (first published 1940) *Farm animals*. Edward Arnold, London, UK

Hane, M., Huber-Eicher, B. and Frohlich, E. (2000) Survey of laying hen husbandry in Switzerland. *World's Poultry Science Journal* **56**: 21-31

Hawk, W. (1910) *Poultry keeping for profit*. Cornwall County Council

Herenda, D. and Jakel, O. (1994) Poultry abattoir survey of carcass condemnation for standard, vegetarian and free range chickens. *Canadian Veterinary Journal* **35**: 293-296 (Abstract)

Howes, H. (1939) *Modern poultry management*. Macmillan, Basingstoke, UK

Huber, H.U. and Fölsch, D.W. (1985a) Hens thrive well on natural light. *Poultry*, **1**: 6-7

Huber, H.U. and Fölsch, D.W. (1985b) The hens' need for light.1985b, *2nd European Symposium. Poultry welfare*. Celle. 292-293

Hughes, B.O. and Duncan, I.J.H. (1972) The influence of strain and environmental factors upon feather pecking and cannibalism in fowl. *British Poultry Science* **13**: 525-547

Hutchinson, J.C.D. (1956) Control of seasonal variation in egg production of hens. *Nature* **177:** 795-796

Jin, L.Z., Ho, Y.W., Abdullah, N. and Jalaludin, S. (1997) Probiotics in poultry: modes of action. *World's Poultry Science Journal* **53:** 351-368

van Kampen, M. (1981) Thermal influences on poultry. In: *Environmental aspects of housing for animal production.* Edited by Clark, J.A., Butterworths, London, UK, pp. 131-147

King, D.F. (1959) Artificial light for growing and laying birds. *Alabama Agricultural Experimental Station Progress Report* 72 (Abstract)

King, D.F. (1961) Effects of increasing, decreasing and constant lighting treatments on growing pullets. *Poultry Science* **40:** 479-484

King, J.R. and Farner, D.S. (1964) Terrestial animals in humid heat: birds. In: *Handbook of physiology.* Edited by Dill, D.B., American Physiological Society, Bethesda, USA, pp. 603

Kleiber, M. (1961) *The fire of life.* Wiley, New York, USA

Kleiber, M. and Doherty, J.E. (1934) The influence of environmental temperature on the utilisation of food energy in baby chicks. *Journal of General Physiology* **17:** 701-726 (Abstract)

Knott, C.I.F. (2000) Personal communication

Korver, D.R. and Klasing, K.C. (1995) Fish oil and fenleuten, a lypoxygenase inhibitor, improve growth of broiler chicks challenged with coccidia. *Poultry Science* **74:** Supplement, 60

Kwakkel, R.P. (1999) Rearing the laying pullet - a multiphasic approach. In: *Recent developments in poultry nutrition* 2. Edited by Wiseman, J. and Garnsworthy, P.C., Nottingham University Press, Nottingham, UK, pp. 227-249

Lampkin, N. (1997) *Organic poultry production.* Report to MAFF, CSA 3699

Larbier, M. and Leclercq, B. (1994) Translated by Wiseman, J., *Nutrition and feeding of poultry.* Nottingham University Press, Nottingham, UK

Leeson, S. and Summers, J.D. (1978) Voluntary food restriction by laying hens mediated through dietary self selection. *British Poultry Science* **19:** 417-424

Leeson, S. and Summers, J.D. (1997) *Commercial poultry nutrition.* University Books, Guelph, Canada

Lewis, P.D., Perry, C.G. and Morris, T.R. (1997) Effect of size and timing of photoperiod increase on age at first egg and subsequent performance of two breeds of laying hen. *British Poultry Science* **38:** 142-150

Lewis, P.D., Morris, T.R. and Perry, G.C. (1999) Light intensity and age at first egg in pullets. *Poultry Science* **78:** 1227-1231

Lewis, P.D., Perry, G.C. and Morris, T.R. (2000) Genetic influences on photoperiodic stimulation of sexual maturity in pullets. *British Poultry Science* **41:** 702-703

Mack, S., Bercovici, D., de Groote, G., Leclercq, B., Lippens, M., Pack, M. Schutte, J.B. and van Cauwenberghe, S. (1999) Ideal protein profile and dietary lysine specification for broiler chickens of 20 to 40 days of age. *British Poultry Science* **40:** 257-265

McDonald, P., Edwards, R.A., Greenhalgh, J.F.D. and Morgan, C.A. (1995) *Animal nutrition*. Longman Scientific and Technical, Harlow, UK

McLeod, M.G. and Dabutha, L.A. (1997) Selection between a "high protein" and a "high energy" diet by Japanese quail in relation to ambient temperature and metabolic weight. *World's Poultry Science Association (UK Branch). Proceedings of Spring meeting*, 43-44

Mead, G.C. (1998) The safety of poultry products: present trends and future developments. *7th Temperton fellowship report*. Harper Adams Agricultural College

Metchnikoff, E. (1908) *Prolongation of life*. Putnam and Sons, New York, USA (Abstract)

Mongin, P. and Saveur, B. (1974) Voluntary food and calcium intake by the laying hen. *British Poultry Science* **15:** 349-359

Morris, T.R. (1968) Light requirements of the fowl. In: *Environmental control in poultry production*. Edited by Carter, T.C., Oliver and Boyd, Edinburgh, UK, pp. 15-39

Morris, T.R. (1986) Nutritional research: past, present and future. In: *Nutrient requirements of poultry and nutritional research*. Edited by Fisher, C. and Boorman, K.N., Butterworths, London, UK, pp. 1-7

Morris, T.R. (1994) Lighting for layers - what we know and what we need to know. *World's Poultry Science Journal* **50:** 283-287

Morris, T.R. (1999) Sexual maturity, lighting and layer performance. Poultry lighting seminar, University of Bristol, 10-12

Morris, T.R. and Fox, S. (1958a) Light and sexual maturity in the domestic fowl. *Nature* **181:** 1453-1454

Morris, T.R. and Fox, S. (1958b) Artificial light and sexual maturity in the fowl. *Nature* **181:** 1522-1523

Morris, T.R. and Owen, M. (1966) The effect of light intensity on egg production. Proceedings XIII World's Poultry Congress, Kiev, pp. 458-461

Morris, T.R., Al-Azzawi, K., Gous, R.M. and Simpson, G.L. (1987) Effects of protein concentration on responses to dietary lysine by chickens. *British Poultry Science* **28:** 185-195

Nurmi, E. and Rantala, M. (1973) New aspects of *Salmonella* infection in broiler production. *Nature* **241:** 210-211

Ofek, I., Mirelman, D. and Sharon, N. (1977) Adherence of *Eschericia coli* to human mucosal cells mediated by mannose receptors. *Nature* **265:** 623-625

Oyofo, B., DeLoach, J.R., Corrier, D.E., Norman, J.O., Zirpin, R.I. and Mollenhauer, H.H. (1989) Prevention of *Salmonella typhimurium* colonisation of broilers with D-mannose. *Poultry Science* **68:** 1357-1360

Parkhurst, R.T. (1928) Artificial light for late hatched pullets. *Eggs,* Scientific Poultry Breeders Association, December 1928, 270-271

Payne, C.G. (1966) Environmental temperature and egg production. In: *The physiology of the domestic fowl*. Edited by Horton-Smith, C. and Amoroso, E.C., Oliver and Boyd, Edinburgh, UK, pp. 235-241

Peisker, M. (1996) Amino acid profiles for poultry. In: *Recent advances in animal nutrition*. Edited by Garnsworthy, P.C., Wiseman, J. and Haresign, W., Nottingham University Press, Nottingham, UK, pp. 57-70

Permin, A. and Nansen, P. (1996) Parasitological problems in organic poultry production. *Kongelige Veterinaer og Landbohojskole* **729:** 91-96 (Abstract)

Prickett, J.D., Robinson, D.R. and Block, K.J. (1982) Enhanced production of IgE and IgG antibodies associated with a diet enriched in eicosapentanoic acid. *Immunology* **46:** 819-826

Renema, R.A., Robinson, F.E., Oosterhoff, H.H., Feddes, J.J.R. and Wilson, J. (2001) Effects of photostimulatory light intensity on ovarian morphology and carcass traits at sexual maturity in modern and antique egg-type pullets. *Poultry Science* **80:** 47-56

Richards, S.A. (1974) Aspects of physical thermoregulation in the fowl. In: *Heat loss from animals and man*. Edited by Monteith, J.L. and Mount, L.E., Butterworths, London, UK, pp. 255-275

Robinson, L. (1948) *Modern poultry husbandry*. Crosby Lockwood, London UK

Robinson, L. (1961) *Modern poultry husbandry*. Crosby Lockwood, London, UK

Romijn, C. and Lokhorst, W. (1966) Heat regulation and energy metabolism in the domestic fowl. In: *Physiology of the domestic fowl*. Edited by Horton-Smith, C. and Amoroso, E.C., Oliver and Boyd, Edinburgh, UK, pp. 211-227

Rose, S.P. and Salah Uddin, M. (1997) The effect of temperature on the response of broiler chickens to lysine balance in the dietary crude protein. *World's Poultry Science Association (UK Branch), Proceedings*, Scarborough, 29-30

Saunders, C.N. (1958) Keratoconjunctivitis in broiler birds. *Veterinary Record* **70:** 117

Sinurat, A.P. and Balnave, D. (1986) Free-choice feeding of broilers at high temperatures. *British Poultry Science* **27:** 577-584

Smith, K., Charles, D.R. and Moorhouse, D. (2000) Nitrogen excretion by farm livestock with respect to land spreading requirements and controlling nitrogen losses to ground and surface waters. Part 2: Pigs and poultry. *Bioresource Technology* **71:** 183-194

Spencer and Charles (1968) Effective environmental control. 2. Light proofing. *World's Poultry Science Journal* **24:** 318

Stopes, C., Duxbury, R. and Graham, R. (2000) UK organic poultry - what do consumers expect? *13th International Scientific Conference of the International Federation of Organic Agriculture Movements*, Basel

Tsiagbe, V.K., Cook, M.E., Harper, A.E. and Sunde, M.L. (1987) Enhanced immune responses in broiler chicks fed methionine supplemented diets. *Poultry Science* **66:** 1147-1154

Wathes, C.M. and Charles, D.R. (1994) *Livestock housing*. CAB International, Wallingford, UK

Whetham, E.O. (1933) Factors modifying egg production with special reference to seasonal changes. *Journal of Agricultural Science* **23:** 383-419

Yeates, N.T.M. (1960) Some problems concerning the mechanism of light action on poultry. *World's Poultry Science Journal* **16:** 324-326

9

POULTRY IN WHOLE FARM SYSTEMS

Background to poultry production practice and scope for adaption to organic production

Chapters 9, 10 and 11 discuss the integration of poultry production into whole farm systems. The Summary and Introduction which follow refer to all three chapters. The references for the three chapters are listed at the end of Chapter 11.

Summary of Chapters 9, 10 and 11

Increased consumer demand for organic egg and chicken meat over the last few years led to considerable market growth. This was fortuitous for the UK poultry industry, as competition in home and European markets for commodity broiler meat and eggs was difficult. Diversification into organic egg and chicken meat production opened up new markets for poultry companies, and this was driven by attractive premiums. It was relatively easy for non-organic free-range egg producers to convert to organic production: they already had land available for grazing. The pullet source could be the same as for non-organic free-range egg production. The main point of difference was the feed.

For the poultry meat industry, however, capital investment in new sites and facilities was needed from the outset. This was because it was not usually possible to convert broiler facilities for organic table chicken production, as land was not available for grazing by the birds. The sector also faced difficulties in producing an organic table chicken of similar live weight to a broiler but at almost twice the age. Restricting the feed intake and growth of broiler hybrids was an ethical issue but alternative hybrids to broilers were not available in the UK. This has been addressed of late and hybrids for use in organic production, eggs as well as meat, are likely to continue to be developed.

It would be unfair to suppose that early organic production was not technically challenging. It was. Nevertheless, the standards permitted

Mark Shepherd (ADAS, Gleadthorpe) contributed information to this chapter rapid growth in the UK organic poultry sector, as they were not too demanding. Market growth would not have been as rapid if organic breeder flocks and pullet rearing, and on-farm feed production, had been features of the early systems. However, they are likely to be features of future systems, and they need to be addressed if markets are to be sustained.

The problem will be how to integrate both existing and new organic poultry enterprises into land-based systems; if, in the future, they cannot be separate from the cropping and manuring of land. The recycling of nutrients between the soil, plants and birds is the crux of organic production.

Poultry in the 1950s were included in farming systems prior to intensification, but not within the requirements of an organic system. Furthermore, technical problems that were resolved by moving egg and chicken meat production indoors now need to be addressed for the outdoor systems of today. For example, methods of control of parasitic disease are needed.

Future systems will be much more technically demanding and there will be upward pressure on the costs of organic poultry production. Thus, the next few years will be difficult and challenging for some producers; in particular organic egg producers as the current UK market supply is thought to be meeting demand. Some organic egg producers will convert to non-organic egg production, particularly if they are unable to produce a reasonable proportion of the feed on-farm, or if they are not able to trade nutrients locally.

Since organic eggs and poultry will be produced increasingly on arable farms there will be a need for arable producers to acquire stockmanship skills, to know of the inputs and outputs of poultry production, and to be aware of flock cycles and of the birds' physiological and nutritional requirements. If production is contracted to poultry companies then there will be a need for poultry personnel to be aware of the constraints of organic cropping systems, crop husbandry and inputs and outputs. Thus, for poultry producers and arable producers alike, there is a great deal of technical progress to be made. This needs to happen soon because the standards and regulations are being driven forward fast.

The purpose of these chapters is to consider the options, limitations and technical difficulties associated with integrating poultry into current UK organic rotations. The studies on which these chapters are based draw on existing knowledge derived from non-organic systems, the results of DEFRA-funded studies on organic poultry production, and tools such as the *MANNER* software package, to provide an assessment of the issues involved.

Current systems of organic poultry production in the UK are described, and this includes information on the inputs and outputs of organic egg and chicken meat production, and typical flock cycles. Differences between organic and non-organic poultry production are highlighted, in particular with reference to feeding and housing.

Practical, economic and market supply constraints of organic poultry production systems are identified. Housing options, that is fixed *versus* mobile, are discussed in the context of their relative merits and disadvantages. For example, mobile houses offer a better means of control of parasitic disease than fixed housing but there are other important considerations, some of which may impact on bird welfare. There may also be limits to the provision of shelter within the paddock when poultry occupy grass/clover leys. Permanent shelter, such as that provided by tree canopies, may interfere with cropping, and so shelter may only be available at the edge of the field. Birds may not have access to this, or the distance from the house to the edge of the field may lessen the attraction. Approaches used in work at Gleadthorpe for providing temporary shelter on pasture are discussed. Some of the shelters used were more effective in attracting chickens to them than others. In addition, there are suggestions on how shelters may be used so as to attract birds to specific areas of a paddock; possibly managing the distribution of droppings outdoors.

Poultry are photoperiodic and so seasonal changes in natural daylength will impact on production unless artificial lighting is used effectively. Discussion focuses on lighting for organic pullets and laying hens as light control is critical and there are suggested light programmes for use at different times of the year. However, there is an example of when it will not be possible to apply light programmes during rearing because of changes in natural daylength. There will be differences throughout Europe in the ability to apply light programmes to pullets and hens as latitude affects natural daylength. For some times of the year producers in southern Europe will be better able to apply light programmes than producers in northern Europe and the reasons for this are discussed. See Chapter 8 for more details on lighting.

If most UK producers decide to avoid rearing during 'difficult' times of year then home organic poultry production may become seasonal. This would not just affect egg production for human consumption, as breeder flocks may be managed similarly to pullets.

The requirements for manure storage and handling are considered in the literature reviewed. Poultry manure is a valuable nutrient source but it is also a potential pollutant and a vector for disease transmission. So as to optimise its use in cropping systems its nutrient value should be determined before land application. This is because breed and feed specification affect manure nutrient contents and examples of this are given. Other factors that may affect manure nutrient content and handling are discussed.

There are recommendations on manure storage and handling for the control of food poisoning bacteria and these are based on work done at Gleadthorpe for the Food Standards Agency (FSA). Batch storage, as recommended for the latter purpose, may also be effective in killing some parasitic worm eggs. This is discussed but there is a need for more information.

Rotational systems using mobile housing may be no more effective in controlling parasitic disease than fixed housing systems, if manure used on land later to be occupied by chickens is a source of infection to the birds. Standards and regulations are not likely to be effective in controlling parasitic disease and so methods of control are needed.

There are concerns about the biosecurity of outdoor systems and the possible risk of chickens being contaminated with *Salmonella* and *Mycoplasma* species through contact with wild birds and animals. Although scratch feeding outdoors may be used as a means of encouraging ranging this is to be avoided, since the feed will attract wild birds and animals into the system. The merits of good all round biosecurity are emphasised in the review, but the standard of biosecurity that can be achieved will inevitably be lower than for indoor systems. In future systems of organic breeder production, *Mycoplasma* infection may increase morbidity and mortality throughout the production chain. Methods of biosecurity should be central to the design of developing organic systems, and standards should allow for this.

There is recognition of the important role that pasture should have in meeting the chickens ranging, nutritional and physiological needs. It is thought that these are not being met by the ryegrass based swards currently predominantly used for poultry. There is considerable scope for developing swards or forages for use in organic and extensive poultry systems but this needs practical research. In the meantime, recommendations are given for the management of ryegrass based pasture.

The effects of outdoor poultry on soil ecology have not been widely studied. However, some populations of diurnally active invertebrates were found to be reduced by integrating poultry into orchards. The invertebrates were a rich nutrient source to the birds, but they were also important in the natural control of crop disease. In integrated systems, the effects of poultry on soil ecology and the control of crop disease will be important.

Using *MANNER*, nutrient budgets were constructed for a number of cropping scenarios, including egg and table chicken production. The rotations used were typical UK organic rotations and they provided some to almost all of the birds' cereal requirements. However, the protein component of the feed was met from external sources. This meant that nitrogen was in excess, although the amount varied depending on the rotation, but phosphorus and potassium were near break even. There is scope to improve nutrient recycling, and other scenarios should be tested including wider rotations that produce some feed protein. The possibility of nutrient trading at a local level should also be considered.

Methods of integrating laying hens and table chickens on a batch basis into current organic rotations are illustrated in these chapters. It is expected that as systems develop (e.g. wider rotations including forages for grazing by chickens) methods of integration will be more complex. Furthermore, there may be complications if breeder and pullet flocks are reared on a seasonal basis.

At several points throughout these reviews the constraints on the future development of the UK organic poultry industry are discussed. Poultry companies may be better placed to investigate technical constraints and to find solutions to them than individual independent producers but this may not always be the case. An individual producer, perhaps having a small flock of birds, may be able to invest unpaid labour so as to find a solution. Such a producer may also be innovative in approach to problems. Thus, to evolve standards, the ideas and strategies of organic producers across the spectrum of production should be considered.

It will be important that organic poultry production meets consumer expectations for bird welfare, environmental issues and food safety. All of these requirements should be taken into account as standards develop. Although the organic poultry market is relatively new there will be future opportunities for designer foods (e.g. organic omega-3 rich eggs and chicken meat) and for added value products. If production does become seasonal as poultry are integrated into cropping systems, either because of light control or forage availability, then acceptable means of storing organic poultry products will be needed. For example, will frozen organic chicken meat be acceptable to consumers or will the demand be only for fresh meat?

Specific issues identified were difficulties in fully closing the system, in terms of feed production for poultry and nutrient sustainability for cropping. This means that there would be value in co-operation for feed production and manure use. The sector bodies could have an important role in bringing together producers at a local level.

Novel production systems such as agroforestry are also thought to have potential; the tree canopy would provide shelter from some predators and microclimates. The poultry would be an additional income to the producer and there may be production benefits associated with poultry manuring the land.

There are many gaps in knowledge and these are discussed; not surprisingly therefore, there are many recommendations for further research.

Introduction to Chapters 9, 10 and 11

The majority of poultry kept in the UK are farmed intensively. There are good biological and economic reasons for this, but it has had the effect of separating much of the poultry production from the land. This is firstly because many of the production units have tended to be businesses established on holdings lacking land. Secondly the personnel, including not only the entrepreneurs but also the supporting work force and professionals, the culture and the skills, have all been different from those applied to the management and cultivation of land.

Many technical and business strengths have resulted from this specialisation of the conventional poultry industry, but a disadvantage is that the exploitation of manure from conventional poultry units in crop rotations, and as a source of plant nutrients, has not always been optimal. There have even been periods when conventional poultry producers have regarded manure disposal as a cost and a nuisance.

These chapters attempt to summarise some of the science and technology relevant to the integration of organic poultry within theoretical rotations. In addition they provide information on the main inputs and outputs of poultry enterprises (laying hens and table chickens), production cycles and husbandry issues. This type of information will be needed by organic arable producers diversifying into poultry production.

The main of points of difference between current non-organic and organic systems of egg production, and between non-organic and organic systems of table bird production, are given in Tables 1 and 2 (Council Directive 1999/74/EC; UKROFS, 2001). This information will be of use to existing non-organic free-range egg or free-range table chicken producers considering conversion to organic production systems.

Poultry production

An understanding of the main inputs and outputs of organic chicken production, and of production cycle lengths is needed before planning the integration of chickens into a whole farm system.

INPUTS AND OUTPUTS

The main physical inputs into organic egg production are pullets and feed. UKROFS standards for organic egg production currently allow UK producers to buy-in conventional pullets up to 18 weeks of age, they are then reared for a six week period according to UKROFS standards, and after this period the eggs may then be sold as organic. The EU and UK derogation for organic pullets is only until 31 December 2003. In view of this, MAFF acted proactively and funded a desk study to examine some of the possible technical difficulties that may arise when rearing pullets in an organic system (OF0192). Some of the technical difficulties identified by MAFF-funded project OF0192 are discussed in later sections of this review and in Chapter 8.

Table 1. MAIN POINTS OF DIFFERENCE BETWEEN NON-ORGANIC FREE-RANGE EGG PRODUCTION SYSTEMS AND ORGANIC EGG PRODUCTION SYSTEMS

Criteria	Production system	
	Non- organic free-range egg production[1]	*Organic egg production[2]*
Maximum age at transfer to laying facilities	None	18 weeks of age
Breed specification	None	Traditional breeds preferred
Maximum house stocking density	9 hens/m^2 (from 01/01/02, note see Directive for derogation)	6 hens/m^2
Flock size	-	3 000
Access to range	-	Daytime access for at least one third of their life
Outdoor area available in rotation/bird	10 m^2	10 m^2
Nest box allowance	7 laying hens per nest or in the case of common nest boxes 120 sq.cm/hen (from 01/01/02)	7 laying hens per nest or in the case of common nest boxes 120 sq.cm/hen
Perch space allowance	15 cm/hen (from 01/01/02)	18 cm/hen

[1]According to Council Directive 1999/74/EC (1999)
[2]According to UKROFS Standards (September, 2001)

An additional requirement for organic egg layers is that the runs must be rested after depopulation for at least two months for adequate disease control, and to allow vegetation to grow back (UKROFS, September 2001). For organic table birds, runs must be rested for a total of at least two months within one calendar year.

Table 2. MAIN POINTS OF DIFFERENCE BETWEEN NON-ORGANIC POULTRY MEAT PRODUCTIONS SYSTEMS AND ORGANIC POULTRY MEAT (CHICKENS ONLY)

Criteria	Production system			
	(a) Intensive broiler	*(b) Extensive*		*(c) Organic[2] table bird*
		Free-range table bird[3]	*Traditional free-range table bird[3]*	
Minimum age at slaughter (days)	None, but generally between 39 and 45	56	81	81
Breed specification	None	None	Slow growing	Slow growing
Maximum house stocking density	34.0 kg live weight/m²	13 birds/m² or a max. of 27.5 kg live weight/m²	12 birds/m² or a max. of 25.0 kg live weight/m²	10 birds/m² with a max. of 21.0 kg live weight/m² (fixed houses)
Flock size	Unlimited	Unlimited	Maximum 4 800 birds	Maximum 4 800 birds
Access to range	Not required	Continuous daytime access required for at least half their life-time	Continuous daytime access required at least from 42 days of age	Daytime access required for one third of their life
Pasture allowance (note for organic area is given as outdoor area available in rotation per bird)	None	1 m² per bird	2 m² per bird	4 m² per bird
Feed specification	None	Finisher contains at least 70% cereals	Finisher contains at least 70% cereals	Finisher contains at least 65% cereals

[2] According to UKROFS Standards (September, 2001)
[3] According to Regulation 2891/93 Article 10 (1999)

MAFF-funded project OF1092 did not consider the effect of rearing system on pullet cost, but this was addressed by Lampkin (1997), in a review of organic poultry production. Table 3 shows the costs of rearing pullets to 20 weeks of age in different production systems as calculated by Lampkin (1997). He noted that at the time of writing typical costs associated with non-organic pullet rearing were £2.50/pullet.

Table 3. REARING COSTS FOR ORGANIC AND CONVERTED CONVENTIONAL PULLETS

	Organic, range-reared (£/pullet)	Converted, range reared (£/pullet)	Converted conventional (£/pullet)
Total rearing costs to 20 weeks of age	4.33	3.97	3.57

Source: Lampkin (1997)

In a typical laying year (20 to 72 weeks of age), an organic layer is expected to eat between 47 and 50 kg of feed. Layers are always fed *ad libitum*. The hens adjust their feed intake during cold or warm weather so as to satisfy their energy demands (e.g. Emmans and Charles, 1977). Thus, birds on free-range tend to eat more than those in controlled environment, and they tend to consume more in winter than in summer.

Hens will use between 50 and 150 kg of water per bird per year. Water usage is affected by drinker type, diet specification, antinutritive factors, feed palatability and thermal environment. In fixed housing, the water supply will be permanant and automatic, but this will not be the case when using mobile housing. It will be necessary to plan the supply of water to the hens, and to ensure that the water supply does not freeze during cold weather. Water restriction is a serious welfare problem and egg mass output will quickly fall if water intake is limited.

Significant outputs from an organic egg enterprise include some 260 to 310 eggs per hen per year and about 40 to 45 kg of fresh droppings per hen per year. Producers will need to consider means of collecting eggs and transporting them without breakage to a suitable egg store. This will be more labour intensive and more difficult when using mobile housing in a *FRFR* system. The requirements for egg storage are discussed below. Depending on whether or not eggs are sold at the farm gate it may be necessary for producers to contract supply a packing station specialising in organic eggs, or to grade and pack eggs according to Egg (Marketing Standards) Regulation 1995 which implements EU regulations on Marketing Standards, as well as UKROFS Standards. A voluntary Code of Practice for the Handling and Storage of Eggs from Farm to Retail Sale is available on DEFRA's website (http://www.defra.gov.uk/foodrin/poultry.pdfs/codesofpractice.pdf).

Depending on the time of year that the hens are depopulated and the laying houses are cleaned out it may be necessary for producers to store the manure prior to planned application in the crop rotation. UKROFS Standards require that storage facilities for livestock manure are of a capacity to preclude the pollution of water by direct discharge, or by run-off and infiltration of the soil. Furthermore, the capacity of the storage facility for livestock manure must exceed the storage capacity required for the longest period of the year in which any application of manure to the land is either inappropriate (in accordance with the Codes of Good Agricultural Practice, published by the Ministry of Agriculture, Fisheries and Food in 1998) or when such application is prohibited.

For non-organic broiler chickens the main physical inputs are chicks, about 5.9 kg of feed per bird and 7.2 kg of water. The main outputs are birds weighing between 2.0 kg and 2.5 kg, and about 3.5 kg per bird of manure. The latter is given for birds grown to 49 days of age, whereas typical market live weights may be achieved as early as 40 days of age. Feed intakes for organic table chickens grown to 81 days of age will be higher, but the values will be affected by breed and feeding strategy. Table 4 shows feed intakes for two flocks of as hatched ISA 657 birds grown in an extensive production system at ADAS Gleadthorpe to 81 and 82 days of age during winter and summer months respectively (DEFRA-funded project OF0153).

Table 4. FEED INTAKES OF AS HATCHED ISA 657 BIRDS GROWN IN AN EXTENSIVE PRODUCTION SYSTEM AT ADAS GLEADTHORPE TO 81/82 DAYS OF AGE (kg/BIRD)

Time of year	Feed intake (kg/bird)
Winter	7.61
Summer	6.64

Source: DEFRA-funded project OF0153
Note: that in both flocks feed intakes are given for birds allowed access to pasture from day 42.

Feed intakes of table chickens to any given age are substantially higher on free-range than they are for intensive broilers because the birds are colder. The effects of temperature on *ad libitum* feed intake in broiler hybrids have been estimated by several teams over the years, including Charles *et al.* (1981), and more recently, with more modern hybrids, by Homidan *et al.* (1996), by Yahav *et al.* (1996). However, all of this work has been in the context of intensive broiler production, and whilst several workers have examined the effects of constant high ambient temperatures, or fluctuating high temperatures, on performance, none have used fluctuating low ambient temperatures.

Brooding and the age at which birds are given access to range will affect flock feed intakes. Chicks are brooded indoors as they have a high thermal requirement (Charles, 1986) and this would not be met by brooding outdoors in the UK. As the birds get older the feathers develop and this provides insulative benefits. Their ability to produce heat and maintain body temperature also increases with age.

Regulation EC 1804/1999 and UKROFS Standards require that birds are given access to range for at least one third of their life. If an 81 day growing period is used then birds must be given daily access to range by at least 60 days of age. It is thought that organic producers would only delay giving birds access to range until 60 days of age in the poorest of weather conditions, as consumers expect organic table chickens to have outdoor access for most of their life. In warm summer months during dry and windless weather it would be feasible to allow birds access to range from as early as 21 days of age, and possibly slightly earlier, but producers must frequently observe the birds so as to check whether or not they are huddling to keep warm.

Breed choice for organic table chicken production is not straightforward. Until recently, only modern broiler hybrid chicks were readily available in the UK. Modern broiler hybrids have a very fast early growth rate and this makes them less suited to organic table chicken production as they exceed typical market live weights of between 2.0 and 2.5 kg well before the minimum slaughter age. If producers choose to use modern broiler hybrids then they would either have to establish a market for heavy organic table chickens, and this may be difficult as the dressed birds are likely to be out of the consumers price range, or to restrict the daily feed intake of broiler hybrids so that the attainment of market live weight is delayed until

81 days of age. For the reasons described below, restriction is likely to be unacceptable.

An experiment at ADAS Gleadthorpe (DEFRA-funded project OF1053) examined growth rates of UK broiler hybrids fed either presumed non-limiting rations (with respect to the Ross 308, commercial broiler), or Label Rouge type rations. As-hatched live weights at days 56 and 81 are shown in Figure 1.

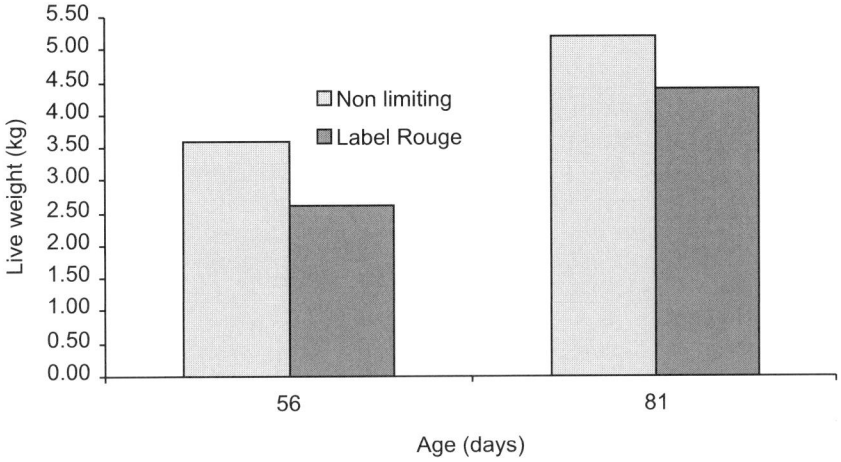

Figure 1. Mean live weights of as-hatched Ross 308 broilers at days 56 and 81 when fed *ad libitum* either presumed non-limiting rations, or Label Rouge type rations and grown in a controlled environment with no access to pasture.

The modern broiler hybrid weighed about 4.4 kg at day 81 when fed *ad libitum* Label Rouge rations. Feed restriction may not be acceptable to consumers of organic produce. Also, if the suggestion by Emmans and Fisher (1986) that animals seek to eat because they seek to grow, and presumably the seeking to grow is driven by the genome seeking to meet its programmed mature body size, then feed restriction may jeopardise bird welfare. Emmans and Fisher (1986) proposed the following equation:

$$DFI = R_1/C_1$$

where, DFI = desired feed intake, kg per day
R_1 = requirements, units per day, for the first limiting resource
C_1 = the feed content, units per kg, of the first limiting resource.

Although it could be argued that birds in extensive systems would have access to pasture, and that intake of pasture is unlimited, fresh grass is very low in protein, lysine and methionine, compared with Label Rouge type rations (Table 5), and moisture contents and crude fibre contents are much higher in grass. A high crude fibre content would limit grass intake as bulkiness would affect gut fill and digestibility. Low lysine and methionine contents in grass, these being the first and second limiting amino acids for growth (Larbier and Leclercq, 1994), would not assist in reducing the birds' desire to feed as described by the Emmans and Fisher (1986) equation.

Table 5. CRUDE PROTEIN, LYSINE AND METHIONINE CONTENTS OF TYPICAL BROILER RATIONS, TYPICAL LABEL ROUGE TYPE RATIONS AND GRASS, GIVEN ON A FRESH BASIS (g/kg)

Feed type	Ration	Dry matter (g/kg)	Crude protein (g/kg)	Lysine (g/kg)	Methionine (g/kg)
Rouge[1]	Starter	873.0	182.0	1.03	0.37
	Grower	871.0	161.0	0.86	0.31
	Finisher	870.0	159.0	0.76	0.29
Grass[2]	Young	200.0	31.2	-	-
	Mature	282.0	28.2	-	-
	Dried	870.0	-	7.1	3.0

Sources:
[1]DEFRA-funded project OF0153
[2]McDonald *et al.,* (1995)

Furthermore, when quantitative feed restriction techniques are applied to birds in communal groups, it is likely that not all birds are able to feed at the same time, and some birds will receive less than their share of feed. Increased competition at feeding may cause aggression and bird damage, and even moderately short periods of feed withdrawal have been shown to cause physiological stress. Karunajeewa (1987), citing the work of Freeman *et al.,* (1980), reported increased plasma corticosterone concentrations two hours after the onset of feed withdrawal in three-week old Light Sussex birds. The birds also developed hypoglycaemia and hyperlipacidaemia. In later work by Freeman *et al.,* (1981), birds had elevated plasma corticosterone concentrations for five weeks following feed restriction early in life, and plasma fatty acid concentrations remained high for eight weeks. This would mean that restricting feed to broiler hybrids in early life would cause physiological stress for at least 43 to 69% of the duration of an 81-day growing period, and feed restriction is thought to be necessary for the majority of the growing period if target live weights of between 2.0 and 2.5 kg are to be reached at day 81. Lewis *et al.,* (1997a) reported that Ross 1 broilers fed so as to achieve a similar live weight at day 83 as ISA 657 hybrids, had feed intakes of only 55% of that of *ad libitum* fed Ross 1 broilers.

It seems that the primary selection criterion for identifying breeds suited for use in an organic production system should be growth rate, and the attainment of a required market live weight no sooner than the minimum slaughter age when feed is not restricted. There are wide differences in growth rates both between commercially available hybrids (mostly as imported stock from France) and between traditional UK breeds. In work at ADAS Gleadthorpe (DEFRA-funded project OF0153) growth rates of several commercial hybrids and traditional breeds fed either presumed non-limiting rations (with respect to the Ross 308, commercial broiler), or Label Rouge type rations were examined. As-hatched live weights achieved at days 56 and 81 are shown in Figures 2 to 5.

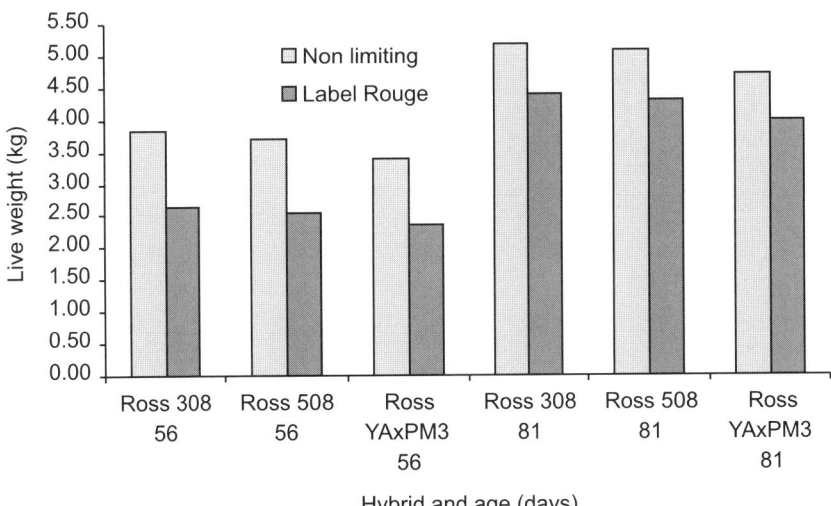

Figure 2. As-hatched live weights at days 56 and 81 for UK commercial broiler hybrids fed either presumed non-limiting rations (with respect to Ross 308), or Label Rouge type rations, when housed in a controlled environment

Figure 3. As-hatched live weights at days 56 and 81 for imported fast-moderate growing hybrids (ISA) fed either presumed non-limiting rations (with respect to Ross 308), or Label Rouge type rations, when housed in a controlled environment

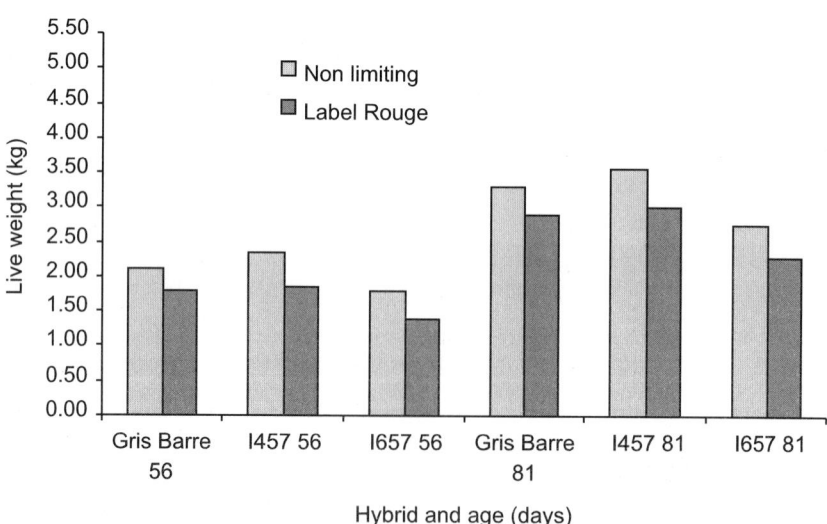

Figure 4. As-hatched live weights at days 56 and 81 for imported moderate-slow growing hybrids (ISA) fed either presumed non-limiting rations (with respect to Ross 308), or Label Rouge type rations, when housed in a controlled environment

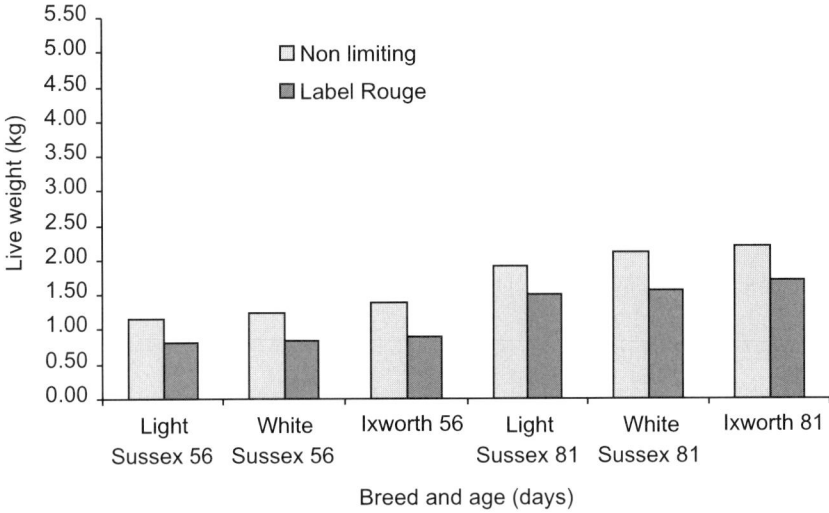

Figure 5. As-hatched live weights at days 56 and 81 for traditional breeds fed either presumed non-limiting rations (with respect to Ross 308), or Label Rouge type rations, when housed in a controlled environment

Differences in growth rates between commercial hybrids are real and deliberate as the breeder companies have developed hybrids suited for particular markets. For example, the phenomenal growth rates of modern broiler hybrids demonstrate the success of broiler breeding companies in selecting for early growth. This trait, and commensurate improvements in feed conversion efficiency, have provided an abundant and relatively cheap supply of poultry meat for UK consumers.

Breed choice and feeding strategy may also affect meat yield, breast meat yield and eating quality characteristics. Gordon (2000) reviewed the effects of length of growing period, breed, management practices and feeding strategies on the quality of extensively produced poultry meat, including organoleptic properties and nutrient content (DEFRA-funded project LS3501). There are also financial implications as chick costs may differ depending on availability and there will be breed differences in feed conversion efficiency.

As the organic table chicken market and other niche markets for extensively produced table chickens continue to expand breed companies may introduce new hybrids that are better suited to UK consumers' requirements. Organic producers considering establishing a table chicken enterprise would be well advised to contact breed companies and the Rare Breeds Survival Trust so as to gather information on breed suitability and chick availability.

Producers will need to consider means of catching and depopulating birds at the end of the growing period. It will be easier to depopulate birds grown in fixed housing as the houses are likely to be more spacious, the facilities for the catchers will be better, and access roads will mean that the lorry can park outside the house. Mobile houses used in a FRFR system will be harder to depopulate and it is likely that the lorry will not be able to park near to the houses. This may mean that the birds have to be loaded into crates and then transported by tractor and trailer to the lorry. The risk of injury to the birds is greater when handling and rehandling, and it is likely to be more stressful to the birds. There is also a high risk that birds may be too cold or too hot when using make-shift transport systems. Depending on the size of the organic table chicken enterprise and the logistics, it may not be possible to use contract catchers. This is because their earnings are related to the number of birds depopulated per unit time. It will be easier to depopulate intensively housed broilers and the number of birds housed on one site will be greater than for an organic unit. Producers should have knowledge of the Wefare of Animals Transport Order (1997).

Producers may contract grow organic table chickens, in which case slaughtering and processing will be organised, but if this is not the case an agreement with a slaughterhouse specialising in organic table chickens will be needed. UKROFS Standards stipulate a maximum journey time between farm and destination of 10 hours duration. The journey time is defined as the time between loading of the first bird and the unloading of the last bird in the consignment. This means that in circumstances requiring a long depopulation and loading time the farm must be located closer to a slaughterhouse than when depopulation is more efficient.

It is beyond the scope of the book to consider the logistics, and financial implications of establishing on-farm slaughtering facilities.

Litter manure output has been estimated as about 3 500 kg per 1000 birds. This will depend on the breed and feeding strategy, drinker type and water spillage, the extent of weather penetration through the popholes, management of the ventilation system and pasture management and the amount of mud brought into the house by the birds. It may be necessary to add litter if the existing litter becomes damp and compacted, so as to prevent hock burn damage and breast blisters. This will increase the amount of litter manure at the end of the growing period. As for layer manure it may be necessary to store the litter manure until there is a sufficient quantity to use in a crop rotation, and until the optimal time of year for manure application.

Production cycles

The production cycle of a batch of non-organic laying hens is typically 60 weeks long. This is composed of a laying period from about 16 to 72 weeks of age, plus a four week turn round time for the cleaning and resting of houses. In organic production systems the turn round time will be determined by: 1) UKROFS Standards, requirement to rest pasture for at least two months; 2) pasture availability, and; 3) the age at which producers aim to provide birds outdoor access. Regulation EC 1804/1999 stipulates that chickens must have access to an open-air run whenever the weather conditions permit and, whenever possible, for at least one third of their lives.

For organic table chickens the production cycle depends upon the weight of bird required by the market. Birds are grown to ages ranging from about 73 days (only if using an approved slow growing strain, but at the time of writing a list of approved strains was not available) to about 81 days plus. The houses are left empty for cleaning for periods varying from a few days to about two weeks or more, depending on circumstances. Thus, there could typically be a maximum of four batches per year (see Chapter 11).

Alternatively, chicks may be brooded in specialist brooding facilities according to UKROFS Standards and then transferred to the free-range facility for the remainder of the growing period. This may enable an extra flock to be produced per annum (see Chapter 11).

In either *RIFR* systems where the available pasture around the house is limited the number of production cycles per year may be fixed by the requirement to rest pastures for a two month period within a calendar year.

Poultry husbandry

HOUSING

Poultry housing has several purposes. Obvious practical functions include keeping the litter, the feed and the birds dry and protected from the elements. It provides a suitable location in which to place the furnishings such as the feeders, drinkers, nests and perches. However housing has more subtle functions. The birds have important, well documented, physiological requirements for the climatic factors temperature and light (e.g. Emmans and Charles, 1977; Charles, 1994; Charles *et al.*, 1994). Lighting is discussed in more detail below, and in Chapter 8, in view of its importance. The housing is intended to provide an environment consistent with these requirements, while at the same time having a ventilation system capable of meeting the requirements of the birds for an adequate air supply of a suitable quality (Charles, 1994).

Detail of factors such as natural ventilation systems and maximum air supply requirements is beyond the present scope, but information and relevant design software is available. See also Charles and Walker (2002).

It is, however, worth mentioning some facets of the effects of housing and ventilation on litter moisture content, since maintaining good litter condition is a crucial aspect of the management of free-range birds, for reasons of health, animal welfare, egg cleanliness and food safety. Tucker and Walker (1992) described the effects of house air psychrometrics on litter moisture content in broiler houses. Condensation occurs on surfaces, or in litter, when the surface or the litter temperature is below the air dew point temperature. The air dew point is defined as the temperature at which relative humidity is 100% at the given air absolute humidity (which is in turn defined as and expressed in kg moisture/kg dry air). The risk of condensation was shown to be high when house air temperature is low and relative humidity is high. Therefore one of the determinants of the ventilation rate requirement of poultry is the removal of respired moisture (Charles *et al.*, 1994; Charles and Walker, 2002). A complication was the subject of a recent study by Edge (2000), who showed that in practice the penetration of wind and rain into popholes makes litter management difficult, particularly with some of the larger popholes specified in recent years, and in housing types having no external verandah acting as a buffer zone between the house and the outside weather.

For chickens on free-range there are designs of housing intended to be static, and surrounded by paddocks around which the birds are rotated, and there are mobile houses which can be moved to wherever the land is to be occupied. For FRFR systems it seems likely that mobile housing may be more appropriate on all but the smallest of farms, due to the likely distance between fields in the rotation.

An advantage of larger static housing systems is the opportunity to introduce mechanisation (or partial mechanisation) for the provision of feed, water and the collection of droppings. This leads to a lower cost per bird than in mobile systems. It is also easier to provide mains services (water and electricity). Controlled environment housing is not possible because of the provision of pop holes. Management of the outside area tends to be more of an issue with static houses (Lampkin, 1997). The area immediately adjacent to the house is utilised to the greatest extent. Hence, there is a need for management practices to be adopted (such as putting straw down in this area) to reduce the amount of mud and moisture that is carried into the sheds in wet weather, since the mud may contaminate fittings. Edge (2000) has demonstrated that the amounts of moisture brought into the house constitute a risk to litter quality. Housing is best configured so that the birds enter the pop holes over a perforated floor area, in order to clean their feet before entering the house (Elson, 1995).

Static housing, as currently used in organic production, varies considerably, and so do the flock sizes housed within the units. Reducing the flock size *per se* will increase capital costs, due to proportionally higher building and equipment costs.

Producers wishing to build new Class 3 buildings for organic production will need planning permission (unless the gross floor area is less than 465 m^2). Criteria of distance from roads, other poultry units and dwelling houses will need to be met.

Stopes *et al.* (2000) concluded from six consumer discussion group studies, each containing seven to eight members, and from interviews with 370 consumers, that consumers expect organic eggs to come from a land based system with mobile housing and small flocks, integrated into the whole farm system.

SOME PARASITES AND DISEASES

Poultry are prone to a range of diseases caused by viruses and by bacteria. Poultry diseases may also be classified as enteric and respiratory. In general it is logical to suppose that intensively housed birds are at greater risk of respiratory diseases while birds on range are at greater risk of enteric diseases.

Viruses may be spread by contact or by aerosol, including windborne spread, depending on the type of virus. Some are fragile outside the host, but others are persistent. They can also be introduced onto a site by visitors, rodents, wild birds, pets, insects and equipment. Bacterial infection may be spread similarly, but it is less likely to be windborne, particularly over long distances. Contaminated feed and water can introduce bacterial contamination. Some bacteria, such as *Mycoplasma,* survive only a short time outside the host, but others survive for long periods, particularly if protected by organic matter (Knott, 2000).

A recently published text book on the important organism *Salmonella* was that of Wray and Wray (2000). They reviewed the role of wild birds, pets and rodents in the transmission of *Salmonella* to chickens. Wild birds were recognised as carriers. Although they are rarely identified as infected, short term carriage occurs. In a sample of 382 dead wild birds only 11 were infected. Of 540 wild birds trapped in Norway 180 were gulls, which were the only birds carrying *Salmonella*. Rodents were considered to constitute a potential risk, including risk of *S. enteriditis* and *S. typhimurium*.

Feeding chickens outdoors is sometimes used as a strategy to encourage ranging, but it is likely to attract wild birds and rodents onto the pasture and because of the risks of *Salmonella* infection it should be avoided. Wire fencing should extend underneath the soil so as to prevent rodents from burrowing, but in FRFR systems this will be labour intensive.

Interest is developing in the possible use of herbal feed supplements for organic chickens, to compensate for the withdrawal of veterinary products. Takeuchi (2000) used a supplement described as "enzyme, fermented soybeans, fructo oligosaccharide, chitosan, herb roots, garlic powders *etc.*" In a test on Cobb broilers grown to 2.58 kg to 2.95 kg (depending on treatment), mortality of controls was 7.8% compared with 1.4% for the feed supplement treatment and 2.8% for a treatment with antibiotics. A statistical analysis was not given. He reported that in Japan 40,000 "French black" organic chickens per year had been grown using the supplement to 2.5 kg in 92 days at 2.0% mortality. A Japanese patent is pending. Whether or not this type of supplement may be used in organic chicken production within the UK will depend upon its ingredients and its manufacturing process. Enzymes are not permitted feed ingredients for organic poultry in the UK.

Herbs may offer some potential for preventing and treating bacterial and parasitic infections in chickens, and it is likely that this will be through the controlled feeding of herb-based products. However, products will need to be approved by the EC and UKROFS before they may be used in feeds for UK organic poultry. The extraction of essential oils and their active ingredients will need to be achieved without chemical means.

Internal parasites afflicting chickens include roundworms and coccidia. The control of parasites and enteric diseases has always been a priority aspect of good husbandry on range. The text books of the traditional period stressed this (e.g. Robinson, 1948). *Ascaridia lineata* (a large intestinal roundworm) was a common parasite of free-range chickens in the traditional period (Barger and Card, 1949), who noted that it was not that uncommon for the intestine of the chicken to be completely blocked at the portion inhabited by *Ascaridia*. *Ascarides* occasionally moved from the intestine into the oviduct and worms were sometimes laid in the hen's egg. The damage done to chickens by *Ascaridia lineata* is greatest in young birds.

Barger and Card (1949) emphasised the importance of preventing puddles on pasture and maintaining dry litter as a means of controlling the development of worm eggs. Cultivation and rotation of paddocks was also of value in reducing the number of infectious eggs to which free-range poultry have access, but except for sunlight *Ascaridia* eggs have considerable resistance to climatic influences. The age at which birds have access to pasture will be important in influencing their susceptibility to disease, but it will also influence the build-up of *Ascaridia* on the land. *Ascaridia* eggs passed out of the birds are not infectious, but under favourable conditions of warmth and moisture they embryonate and become infectious in 10 to 16 days (Barger and Card, 1949).

During the period of embryonation a small, coiled worm develops within the *Ascarid* egg and if it is eaten by a chicken the young worm is liberated from the shell and it begins to develop in the intestinal tract of the host (Barger and Card, 1949). A period of 50 to 60 days is required for the young worms to attain maturity, at which time the female worms are capable of laying eggs and starting the cycle over again (*loc. sit.*). Thus, in the case of meat birds, if chicks are brooded and kept indoors until 28 days of age, but then given access to pasture that has previously been used by chickens and infected with *Ascaridia*, there may be sufficient time for the multiplication of *Ascaridia* prior to slaughter. Delaying the age at which birds are given access to pasture may mean that the pasture used by young birds does not become increasingly contaminated. In future organic egg production systems requiring organically reared pullets, the age at which birds are given access to pasture will be important.

Barger and Card (1949) reported that *Heterakis gallinae* was a common caecal worm in free-range chickens of the traditional period. It was frequently found in large numbers in the caeca, particularly in young birds. *Heterakis* eggs excreted in the faeces may become infectious under conditions of sufficient warmth and moisture in about seven to 14 days. The eggs are not readily destroyed by climatic influences and work during the traditional period found that fully developed worm eggs may persist in the soil for more than eight months (Barger and Card, 1949 citing Graybill 1921). About 65 days are required for the entire cycle from egg to mature worm to be completed. *Heterakis* also plays a role in the transmission of Blackhead (infectious enterohepatitis caused by *Histomonas meleagridis*). As for *Ascaridia*, it is important to prevent puddles developing on pasture and for the litter to be dry. Treatment for *Ascaridia* and *Heterakis* would usually be with anthelmintics under veterinary supervision, and for organic production according to regulatory requirements and standards.

Coccidiosis, a disease caused by *Eimeria* (a protozoan parasite), occurs wherever poultry are raised (Barger and Card, 1949). It has devastating effects in chickens and causes high rates of mortality, especially in younger birds. There are eight species of *Eimeria* capable of infecting chickens, but two are particularly pathogenic, namely *Eimeria tenella* and *Eimeria necatrix*. A vaccine containing all eight species of *Eimeria* is available and it is understood that it may be used in organic poultry production. The efficacy of allopathic treatments for coccidia has not been reviewed.

The sporulated oocyst is the infective stage, and sporulation needs warmth, moisture and oxygen (Jordan and Pattison, 1996). They described wet litter as "ideal sporulation conditions", although the optimum temperature for the process was given as 25 to 30°C. The importance of maintaining dry and friable litter within the house for disease control cannot be over-emphasised.

Bray and Lancaster (1992) studied the effects of factors such as distance from the house, and the degree of faecal contamination of the land, on parasitic status of conventional laying hens. They sampled soil for coccidial oocysts and worm eggs on four Welsh free-range farms of 4 000 birds each at 1 000 birds/ha. They found few coccidial oocysts on these particular sites and abandoned that part of the investigation. However, it is likely that in-feed coccidiostats were used during rearing and this would have reduced the birds' susceptibility to *Eimeria*. Embryonated worm eggs were found and the numbers decreased curvilinearly with distance from the houses. Soil samples contained *Ascaridia* (82%), *Heterakis* (16%) and *Capillaria* (3%). The authors deemed the land to be fowl sick.

Robinson (1948) noted that it had been estimated that 20% of the national flock of his time was infested with the large round worm, so the risk of worms to birds on range, and the importance of rotation in controlling them, should not be underestimated. The traditional authors recommending liming to prevent land becoming fowl sick, though their definition of fowl sick is not clear. They recommended clean houses and clean land.

In Denmark, between 1991 and 1996 egg production from free-range/organic systems increased from 0% to 5% (Permin *et al.,* 1999, citing Ambrosen, 1996). During the same period complaints about the occurrence of *Ascaridia galli* in Danish eggs appeared to increase. Permin and Nansen (1996, cited by Permin *et al.,* 1999) suggested that the change of production system would lead to a renewed prevalence of helminth infections in poultry.

The prevalence of gastrointestinal helminths between October 1994 and October 1995 in different poultry production systems in Denmark was studied by Permin *et al.,* (1999). For laying hens, the production systems studied were free-range/organic, deep litter, cage and backyard systems. Except for the backyard system, all hens were about 40 weeks of age at sampling.

Helminths were common in free-range/organic systems (*Ascardia galli* 63.8%, *Heterakis gallinarum* 72.5%, *Capillaria obsignata* 53.6%, *Capillaria anatis* 31.9% and *Capillaria caudinflata* 1.5%) and in deep litter systems (*Ascaridia galli* 41.9%, *Heterakis gallinarum* 19.4% and *Capillaria obsignata* 51.6%) and backyard systems (*Ascardia galli* 37.5%, *Heterakis gallinarum* 68.6%, *Capillaria obsignata* 50.0%, *Capillaria anatis* 56.3% and *Capillaria caudinflata* 6.3%)(Permin *et al.,* 1999).

In cages and broiler parent systems helminths were rarely found (in cages: *Ascaridia galli* 5% and *Raillietina cesticillus* or *Choanotaenia infundibulum* 3.3%, and in broiler/parent systems: *Capillaria obsignata* 1.6%) (Permin *et al.,* 1999).

Heterakis gallinarum was the most common gastrointestinal helminth found in the study, followed by *Ascaridia galli* and then *Capillaria obsignata* (*loc. sit.*).   The life cycle of *Heterakis gallinarum* is direct, but earthworms have a role in transmitting infected eggs (Permin *et al.,* 1999 citing Lund *et al.,* 1966).   This may explain why the prevalence of *Heterakis gallinarum* was higher in free-range/organic systems and backyard systems than in deep litter systems.

Ascaridia galli has a direct life cycle and it is ingested from faeces contaminated areas (Permin *et al., 1999* citing Ackert, 1931).   Thus, in non-cage systems where birds are not separated from their faeces the risk of ingestion is high.

There is some evidence that the amount of protein in the diet has an effect on the establishment of *A. galli* infection in the gut of laying hens kept under free-range conditions (Permin *et al.,* 1998).   Hens given 140 g crude protein/kg of feed had a lower mean worm burden of adult *A. galli* worms than hens given 180 g crude protein/kg of feed.   However, the number of eggs per gram of faeces was not affected by dietary crude protein content.   The mechanism by which diet composition affects worm burdens is not known.   It is suggested that this topic should be given more consideration as organic poultry feeds differ in composition and specification to non-organic feeds.

Capillaria obsignata has a direct life cycle (Permin *et al.,* 1999 citing Wakelin, 1965) with highly resistant eggs.   The eggs develop into infective stages after only 6 to 8 days at room temperature and they remain infective for 14 days at -12 °C. Thus, prolonged severe ground frosts would be needed in order to kill outdoor infective eggs.

Capillaria anatis was only identified in free-range/organic and backyard systems. Soulsby (1982 cited by Permin *et al.,* 1999) described the life cycle of *Capillaria anatis* as probably being direct, but Permin *et al.,* (1999) suggested that intermediate hosts may possibly be needed to maintain the life cycle because confinement eliminated infection.

The findings of Permin *et al.,* (1999) described above were similar to those of Madsen (1952, cited by Permin *et al.,* 1999) who investigated black geese and partridges. Permin *et al.,* (1999) suggested that game birds may be a source of infection for free-range/organic chickens.

Pasture management and rotation will be important in controlling the build-up of parasites on land used by poultry. Monitoring faecal worm counts and periodically sampling birds to establish the worm burdens within the gut will also be important. A veterinary health plan is a requirement of organic poultry production, but veterinary advice should also be sought on the age at which young birds should be given access to pasture and frequency of rotating paddocks, as this may differ according to the parasitic status of the land.

In organic systems, chicken manure is a valuable source of plant nutrients for crops. However, the use of chicken manure on land later occupied by chickens should be considered with care, as the risk of parasitic infection may be high. This suggests that chicken manure should be used on land at a stage in the rotation giving the greatest gap between application and subsequent occupation by birds. This approach is restrictive, and a better approach would be to reduce manure parasite burdens prior to manure application on land. The manure may then be used at a stage within the rotation where its nutrient value is best utilised by the crop.

Composting of chicken manure before application on land may be a means of reducing contamination and parasite build-up in FRFR systems. In work done at Gleadthorpe, unturned broiler litter reached a temperature of between 50°C and 55°C in the main heap between days 10 and 30 of storage (FSA-funded project BOS003). Exposure through time to the above temperatures may kill most or all of the parasites present in the manure.

For example, exposure to temperatures of between 45°C and 50°C for about one day are lethal to oocysts (Gordon and Jordan, 1982). Another benefit associated with storing poultry manure before land application is that pathogens important in human food poisoning (*Salmonella* and *E. coli*) are killed (FSA-funded project BOS003).

Poultry are also prone to external parasites including fleas and mites, in particular red mite. A DEFRA-funded project (AW02024) managed by researchers from Gleadthorpe studied changes through time in the population of red mite (*Dermanyssus gallinae*) in a commercial barn egg production system. The red mite population increased noticeably about four months after housing the pullets and it peaked by five months. There were high numbers of all life stages of the mite (larvae, protonymph, dentonymph, adult males and adult females). This indicated that the population was increasing rather than static; adults would have been dominant in a static population. The study found that the feed troughs, stairs, and egg belts were highly populated by the mites, with fewer mites on the drinkers and nest boxes. Other workers have reported that red mites prefer to hide during the day in crevices close to the hens' roosting sites (Nordenfors and Hoglund, 2000).

In non-organic egg production red mite is a problem and chemical treatments do not provide complete control. Unless there are natural means of preventing red mite infestation in organic production (e.g. aromatic herb oils), red mite will be a problem.

System design has a part to play in the control of red mite. The mites inhabit cracks and crevices and this makes them difficult to target. Joints between wood and wood, or between wood and metal fixtures and fittings, are common hiding places. This suggests that moulded feeders, perches, and nest boxes with minimum joints would be the best solution.

UTILISATION OF STRAW

Straw bedding is used in organic poultry production because woodshavings produced as a by-product of the housing industry are treated with preservatives that are not permitted within Regulation EC 1804/1999. The integration of chickens within whole farm rotations would enable the on-farm use of straw and its recycling when litter manure is applied to the land. However, it would be very important to monitor straw quality in order to avoid the use of mouldy material. Moulds increase the risk of mycotoxin infection in poultry, resulting in disease. For example, *Apergillus* infection can damage lung structure according to information reviewed by *Leeson et al.,* (1995).

The ability to produce sufficient straw on-farm for use in an organic chicken enterprise is considered in a section below.

Egg storage

Eggs in the oviduct are at body temperature (41°C) and cool rapidly to ambient temperature after laying. It is important to cool them as quickly as possible, and to keep them cool, in order to preserve their internal quality, and to slow the growth of micro-organisms.

After lay eggs lose weight over time by the loss of water through the porous shell. Their albumen quality deteriorates by the loss of carbon dioxide through the shell and the consequent increase in albumen pH. Both of these processes are slowed by cool storage and the weight loss is also slowed at the appropriate relative humidity. It is usual to recommend target storage temperatures of 12°C to 15°C for short term on farm storage (MAFF, 1983). Since it is not always possible to financially justify refrigerated storage on farm a number of practical designs for simple egg stores have been produced by organisations such as ADAS. Insulated, windowless rooms, detached from the poultry house, are used, and some have power ventilation for overnight air cooling.

There may be logistical problems associated with transporting eggs from mobile houses to the egg store without causing shell cracks if the transport is over rough terrain. Eggs cracked during transit may dirty neighbouring eggs and this may lead to further losses due to downgrading.

Food safety

No consideration of the place of poultry in the rotation would be complete without stressing the need to consider food safety. Due diligence suggests that pending research on the subject, crops which may be used uncooked for human consumption, such as salad crops and some vegetables, should perhaps not feature in the rotation with organic poultry.

Findings of FSA-funded project BOS003 have been used to produce recommendations aimed at minimising the risk of pathogen transfer from non-organic poultry manures to humans and these are given below.

Chicken manures should be batch stored for at least three months and no new manure should be added to the heap. If manures are actively managed (e.g. by turning and mixing) then one month storage should be sufficient. Ideally manure should be applied to land before lower risk crops such as cereals. For fresh manure, it must not be applied to land where ready-to-eat crops will be grown unless there is at least six months between application and harvest. At least four months should be left between livestock occupying a field and the subsequent harvest of a ready-to-eat crop. For stored manure, there should be at least two months between application to land and the subsequent harvest of a ready-to-eat crop. Lastly, neither fresh manure nor stored manure should be added directly to a ready-to-eat crop.

Arguably research on the potential for contamination of any crop is needed before organic poultry manures can be used with confidence in rotations. The points made by Wray and Wray (2000) given above suggest that free-range systems may be at greater risk from contamination by rodents and wild birds than the best biosecure indoor systems. The microbiological and practical aspects of food safety in poultry production have been well documented, and the following highlights are taken from a recent authoritative review by Mead (1998).

He considered that the most important food safety issues for the poultry industry were related to contamination by *Salmonella* and *Campylobacter* species. *Eschericia coli* 0157 had not been found on poultry meat despite the ability of the organism to colonise the alimentary tract of young chicks in experimental studies. Application of the Hazard Analysis Critical Control Point (HACCP) principle throughout the food chain was considered the most systematic and effective approach to food safety practice. He noted that the rigourous biosecurity measures which can be imposed on intensive systems cannot be imposed on free-range, though once *Salmonella* and *Campylobacter* gain entry their spread is probably more rapid in indoor systems.

Influence of weather on range usage by poultry and the provision of shelter

The outdoor temperature in Britain is seldom within the optimum range for egg production as published by Emmans and Charles (1977), based on the earlier egg production response work of authors such as Payne (1966) and calorimetric work such as that of Romijn and Lokhorst (1966). Furthermore, there are occasions when wind speed and rain increase the environmental heat demand to which the birds are exposed. The physical principles governing the rate of heat loss from animals as affected by environmental variables were reviewed by Clark and McArthur (1994), and the special considerations of the physics of heat loss through feathers as affected by environmental factors were considered in depth by Wathes and Clark (1981a; 1981b).

Although optimum temperatures are rarely likely to be provided while the birds are outside the houses, range shelter from the elements in the form of wind breaks would be of benefit to both bird welfare and to performance. Shelter engineering is a matter of the choice of the porosity and the height of shelter materials or wind breaks, in order to optimise the degree of reduction of wind speed over the desired distance downwind of the shelter. Owen (1994) has provided some recommendations which could be applied to natural or artificial shelters. Graphical quantification of leeward wind velocities as functions of shelter type and porosity were provided.

In *FRFR* systems it will be more labour intensive to move and relocate windbreaks as chickens are rotated around the farm. At a new location the optimal position for windbreaks will need to be determined. Windbreaks may be needed to prevent excessive wind and rain penetration through popholes, particularly on exposed sites and when popholes are located on the side of the prevailing wind. It may be more difficult to use natural shelters except on the boundaries of the field (e.g. hedges) as permanent natural shelters in the pasture will intefere with the subsequent use of the land. Temporary natural shelters in the form of conifer wig-wams have been used successfully in work at ADAS Gleadthorpe on extensive table chickens. The birds found the structures attractive and so they were well-used (Figure 6). A disadvantage was that because the conifers were cut they quickly died and this meant that they needed replacing after a few months. They also had to be secured to the ground to prevent them from being blown over. However, a major advantage is that a structure such as a conifer wig-wam may be moved around the paddock at regular intervals so as to prevent poaching.

Figure 6. Conifer wig-wam used at ADAS Gleadthorpe (DEFRA-funded project OF0153)

Regional climatic information, which could conceivably help in the prediction of the effects of local climate on the interactions between poultry and the rotation, is available. Example sources are Smith (1984) and Wheeler and Mayes (1997). Temperature, wind speed and precipitation are probably the key climatic variables. Table 6 uses these reference sources to summarise the factors affecting some climatic features relevant to outdoor poultry husbandry in Britain, as judged by the bird behavioural observations from DEFRA-funded project OF0153 and from the thermal response literature reviewed below. These may give some indication of the points to be considered in assessing the relative suitabilities of geographical locations. The publications quoted quantified the effects of several of the factors.

Table 6.   FACTORS AFFECTING ASPECTS OF THE BRITISH CLIMATE

Climatic feature	Factors affecting
Temperature	Latitude, altitude, proximity to the coast
Rainfall	Altitude, location relative to hill rain shadows
Sunshine	Latitude, altitude, proximity to the coast, proximity to the south coast
Wind	Distance from south west coast, shelter by uplands

Sources: Smith (1984) and Wheeler and Mayes (1997)

SUMMARY OF ELECTRONIC DETECTION OF THE BIRDS DURING THEIR NATURAL BEHAVIOUR

Chickens range more in warm and non-windy weather than in cold and windy weather (DEFRA-funded project OF0153). The relationships between range usage and weather conditions as determined in DEFRA-funded project OF0153 are given below for free-range female Ross 308 birds and for free-range as hatched ISA 657 birds.

Number of detections on range for female Ross 308 birds = 2.40*minimum ambient temperature (°C) + 20.5*maximum ambient temperature (°C) - 0.441*windrun (km) + 4.0*rainfall (mm) - 85

where: $r^2 = 0.44$, $p<0.05$

Note: determined for chicks brooded in a controlled environment and transferred to range facilities at day 28.

Number of detections on range for as hatched ISA 657 birds = 67.9 - 10.8*minimum ambient temperature (°C) + 4.63*maximum ambient temperature (°C) - 0.061*windrun (km) + 2.37*rainfall (mm)

where: $r^2 = 0.39$, $p<0.05$

Note: determined for chicks brooded in a controlled environment and transferred to range facilities at day 42.

The multiple regressions given above are relevant for the range of weather conditions experienced by the birds during the experiment at ADAS Gleadthorpe, these conditions are summarised in Tables 7 and 8.

The number of detections on range for as hatched ISA 657 birds have been calculated for a range of weather conditions and the results are shown in Table 9. Similar calculations have not been done for female Ross 308 birds as the regression equation given above is only relevant for range usage up to 56 days of age.

Table 7. WEATHER CONDITIONS EXPERIENCED BY FEMALE ROSS 308 BIRDS IN EXPERIMENT 3 AT ADAS GLEADTHORPE (DATES WHEN GIVEN ACCESS TO RANGE WHERE BETWEEN 29/02/00 AND 23/03/00)

Range	Weather conditions			
	Min. ambient temperature (°C)	*Max. ambient temperature (°C)*	*Rainfall (mm)*	*Windrun (km)*
Minimum	-4.6	5.4	0.0	122
Maximum	8.6	12.7	3.4	540

Source: DEFRA-funded project OF0153

Table 8. WEATHER CONDITIONS EXPERIENCED BY AS HATCHED ISA 657 BIRDS IN EXPERIMENT 3 AT ADAS GLEADTHORPE (DATES WHEN GIVEN ACCESS TO RANGE WHERE BETWEEN 14/03/00 AND 17/04/00)

Range	Weather conditions			
	Min. ambient temperature (°C)	*Max. ambient temperature (°C)*	*Rainfall (mm)*	*Windrun (km)*
Minimum	-6.4	6.6	0.0	6
Maximum	11.6	15.7	5.2	369

Source: DEFRA-funded project OF0153

Table 9. CALCULATED NUMBER OF DETECTIONS ON RANGE FOR AS HATCHED ISA 657 BIRDS AS AFFECTED BY WEATHER CONDITIONS

Weather conditions				
Min. ambient temperature (°C)	*Max. ambient temperature (°C)*	*Rainfall (mm)*	*Windrun (km)*	*Calculated number of detections on range*
0	8	8	0	104
0	8	60	0	101
0	8	120	0	98
0	8	240	0	90
0	8	360	0	83
2	14	8	0	111
2	14	60	0	107
2	14	120	0	104
2	14	240	0	96
2	14	360	0	89

Source: DEFRA-funded project OF0153
Note: determined for chicks brooded in a controlled environment and transferred to range facilities at day 42.

The calculated number of detections on range increased as the maximum and minimum ambient temperatures increased, but increasing daily windruns reduced the calculated number of detections.

Subsequent work at ADAS Gleadthorpe found that ranging activity was greater when birds were provided with early access to range (DEFRA-funded project OF1053). ISA 657 chicks brooded in a controlled environment house until 42 days of age ranged less when given access to pasture from 43 days of age than chicks brooded in free-range facilities with access to pasture from 21 days of age.

In the latter study, the effects of providing artificial and natural shelter outdoors *versus* pasture only on pasture usage were examined. Artificial shelter comprised a 1 m high strip of porous windbreak* running centrally down a 25.5 m paddock. The length of the porous windbreak was 12.3 m and the start of the windbreak was 2.5 m from the pophole. In addition, there was overhead shelter comprised of windbreak measuring 2 m long x 1 m wide and this was supported on wooden posts at a height of 1 m above the pasture. Natural shelter was in the form of a 2 m high conifer wig-wam (Figure 6) midway down one side of the paddock and straw bales midway down the other side.

In paddocks that did not have shelter, the birds were most often detected near to the pophole. The conifer wig-wams provided in the enriched pastures were very attractive to the birds, and especially when chicks were brooded in the range facilities with early access to range. In the latter treatment the highest number of detections was at the conifer wig-wam rather than at the pophole. There was very little use made of the artificial shelter, namely the windbreak or the overhead shelter. Overall range usage tended to be better on warmer, non-windy, dry days and this confirms findings from the previous study.

In a later study, it was of interest to see whether or not chickens still used a conifer wig-wam when it was moved at frequent intervals to a new position within the paddock. If they did this would help in paddock management, as usage may be more even than when shelter is fixed or not available. The conifer wig-wam was on the left-hand side of the paddock when the chickens were first given access to pasture, and then it was moved at weekly intervals around the paddock so that by the end of the study it was on the right-hand side. This is illustrated in Figure 7.

---

*Galebreaker Products, Farmplan House, Rank Xerox Business Park, Mitcheldean, Gloucesteshire, GL17 0SN

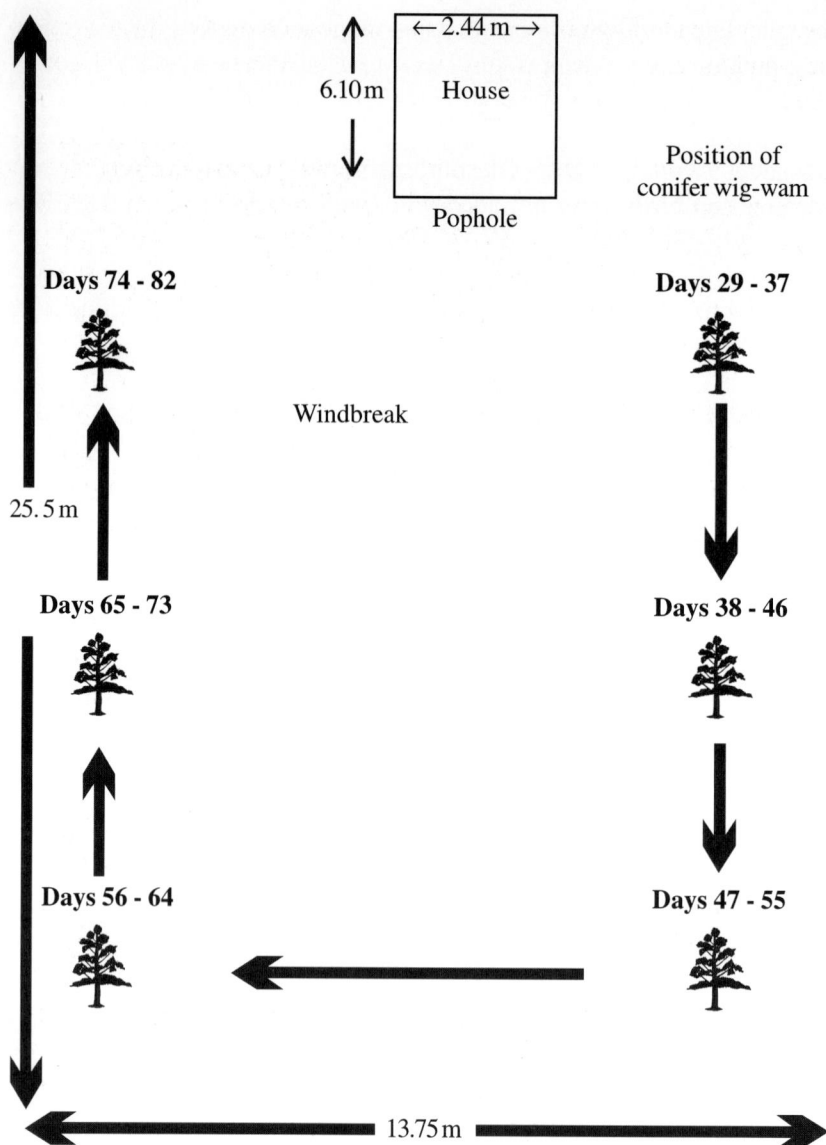

Figure 7. Schedule for the rotation of a conifer wig-wam in all paddocks

Chickens were attracted to the conifer wig-wam regardless of its location. This is shown in Figures 8 and 9 below. In Figure 8, the wig-wam is on the left-hand side of paddock (as viewed by the birds when leaving the house through the pophole) and in Figure 9 the wig-wam is on the right-hand side of the paddock.

Figure 8. Conifer wig-wam on the left-hand side of the paddock as viewed by the birds when leaving the house through the pophole

Figure 9. Conifer wig-wam on the right-hand side of the paddock as viewed by the chickens when leaving the house through the pophole

Being able to manage chickens so as to attract them to different areas of a paddock would be useful for manuring and disease control. In production systems using fixed houses there are limits to the number of options for paddock rotation. The paddocks may be long and thin; extending from the house. If birds remain close to the house then the land will become poached, and manure and parasites will build-up. A row of conifer wig-wams running crossways through the paddock and moved down the paddock away from the house at frequent intervals during the production period may encourage birds to range down the paddock.

The proportion of droppings deposited outdoors is not known, but it suggested that even for good-rangers at least 60% of droppings will be indoors and the remainder will be outdoors. Chickens are fed indoors and feeding invokes the gastro-cholic reflex. Thus, droppings are expected to accumulate close to the site of feeding and this is verified by darker and less friable litter in areas close to the feeders.

Broilers receiving moderate photoperiods such as 12 hours to 16 hours are meal eaters (e.g.Savory, 1976). Peak feed intakes occur at the start and end of the daily light period. As the birds age, the pattern of intake alters so that most feed is consumed later in the day (Siegel and Guhl, 1956). This is probably related to egg formation in the hen, which occurs during the night-time.

In commercial systems chickens range most at the start and end of day, possibly when the prevailing light quality is attractive to the birds. If feeding occurs prior to or after this, then the droppings are likely to be deposited in the house. Droppings deposited outdoors are likely to be in the intervals between meals, and the droppings will be concentrated in the most used areas. For example, in the work at Gleadthorpe, droppings outdoors were most often seen close to the house or near to and under the conifer wig-wam.

Droppings are a disease risk, and parasitic eggs may be ingested through pecking activity. This is one reason why cage systems for egg production were developed; i.e. the system separates the hens from the droppings. In extensive systems of egg production a droppings pit is used indoors to achieve this. Outdoors, the design of the paddock area immediate to the house, and the soil type are likely to be important. A slatted verandah is used in some commercial systems as this separates the birds from the droppings, and it cleans the birds' feet before entering the house. Ideally, there would be a drainage system installed underneath the verandah so that when the verandah is washed the water drains freely and the land does not remain water logged. Parasitic eggs need moisture and so heavy clay soils which allow water to pool are not suited for chicken production.

Chain harrowing may help to distribute droppings more evenly on the land and this may expose parasitic eggs to sunlight, which is effective in killing some parasitic eggs.

The work at Gleadthorpe has quantified some of the effects of weather and shelter on pasture usage by young chickens. There are many possible variations on paddock design and shelter provision and these should be studied with the aim of enhancing pasture use, the spread of droppings on land, shelter from predators and shelter from thermal discomfort. Strategies used in game bird production should be assessed and their suitability for use in production systems for chickens studied. In future systems of organic chicken production, ryegrass based swards are likely to be less important. Strips of forages may be used to provide shelter and nutrient value to the birds and this may reduce the reliance on bought-in feedstuffs. Pasture types and pasture management are discussed in more detail in Chapter 7.

Poultry manurial values and poultry nitrogen balances

The nitrogenous excretion of birds is mainly in the form of insoluble uric acid (Sturkie, 1976). The composition of the excreta of the chicken was reviewed and analysed by Squance (1966), whose results are given in Table 10. His review suggested that urinary nitrogen is mainly composed of uric acid and ammonia, while faecal nitrogen includes albuminoids, digestive enzymes and sloughed off gut lining tissue (presumably proteins). For anatomical reasons it is difficult to separate urine from faeces in the chicken, except by surgical means.

Uric acid ($C_5H_4N_3O_4$) has a structure based on two carbon rings incorporating four NH groups (Lehninger *et al.*, 1993). Despite its relative insolubility, once uric acid is voided it is readily converted to ammonia (NH_3) by micro-organisms. In a review Carlile (1984), quoting several earlier reviews, listed 19 species of bacteria, 22 species of fungi and 3 of

Table 10. COMPOSITION OF THE EXCRETA OF HENS FED A DIET CONTAINING 17 g/kg CRUDE PROTEIN; NITROGEN (N) IN THE CONSTITUENTS IS GIVEN AS PERCENTAGES OF THE TOTAL N

Total N (mg/g dry matter)	Uric N (%)	Ammonia N (%)	Urea N (%)	Creatinine N (%)
77.3	51.0	2.6	1.0	1.5

Source: Squance (1966)

actinomycetes which are capable of degrading uric acid, sometimes operating in succession or in groups. They include both aerobes and anaerobes, though the aerobic population may be more significant in ammonia release. The aerobic degradation of uric acid takes place in several steps, each with its own enzyme, through allantoin, ureidoglycolate, glycolate plus urea, and then finally urease converts these to ammonia and carbon dioxide. As Valentine (1964) realised in the case of the ammonia in poultry house air, it is very soluble and tends to occcur as ammonium hydroxide in solution (NH_4OH).

Nicholson *et al.,* (1996) reported the plant nutrient composition of non-organic poultry manures in England and Wales, including free-range layer manures and broiler manures (Table 11). The nutrient composition of organic layer manure is at present expected to be similar to those given by Nicholson *et al.,* (1996) for non-organic free-range layers, provided that the feed specifications and protein digestibilities are similar between systems. Modern layer hybrids are used in organic and non-organic systems of egg production. However, in the future this may not be the case as there may be discrepancies in protein and amino acid contents between organic and non-organic layer feeds, and between breed types used.

Regulation EC 1804/1999 does not permit the use of synthetic amino acids in organic poultry feeds and although there is a short-term derogation in force for UK organic poultry feeds, a long-term derogation is not likely. Thus, it may be necessary to increase the dietary crude protein content of feeds for organic layers in order to achieve adequate dietary concentrations of essential amino acids. In these circumstances the excretion of nitrogen would be expected to increase. For layers a recent version of the ADAS

Table 11. TOTAL NITROGEN, AMMONIUM-N AND URIC ACID-N MEASURED IN MANURE SAMPLED FROM NON-ORGANIC FREE-RANGE LAYERS

	Non-organic free-range layer manure
Dry matter (%)	57.9
Total N (%dw)	5.5
Total N (kg/t fw)	34
NH_4-N (%dw)	1.1
NH_4-N (kg/t fw)	5.0
Uric acid-N (%dw)	1.8
Uric acid-N (kg/t fw)	11.0

Source: Nicholson *et al.,* (1996)

Hen biological model (originally published by Charles, 1984) has been used to calculate the effect on nitrogen excretion of increasing dietary crude protein content whilst keeping the concentrations of essential amino acids constant (kg per bird per year, Table 12).

Table 12. EFFECT OF DIETARY CRUDE PROTEIN CONCENTRATION ON NITROGEN EXCRETION IN LAYING HENS (KG N PER BIRD PER YEAR)

Dietary crude protein content (g/kg)	*Nitrogen excretion (kg N/bird.year)*
155	0.74
165	0.81
175	0.88
185	0.95
195	1.02

Source: ADAS
Note: calculations were made for hens housed in a thermoneutral environment. Cold temperatures as experienced by organic hens may increase nitrogen excretion because of temperature effects on feed intake.

Differences in protein digestibility and in antinutritive factors between feed ingredients may also affect nitrogen excretion. Soya is widely used as the main protein source in non-organic poultry feeds, but issues of traceability make soya a less desirable ingredient for organic feeds. Home grown protein sources may have lower protein digestibilities than soya, and whilst cooking improves the digestibility of soya through the degradation of trypsin inhibitor, it may not be possible to achieve similar improvements in digestibility in other ingredients. The use of home grown protein sources in feeds for organic poultry is discussed in Chapter 3.

The nitrogen value of organic table chicken manure may be expected to differ from that of non-organic broiler manure. This may be due to breed differences in feed conversion efficiency throughout the growing period, and to differences in the dietary nitrogen content between non-organic broiler rations and organic table chicken rations. Regulation EC1804/1999 and UKROFS Standards require the finisher ration to be comprised of cereals at a proportion of 65% of the ingredients. The high cereal component of the ration dilutes the dietary nitrogen content of the organic ration, compared with non-organic broiler rations.

DEFRA-funded projects OF0153 and OF1063 have provided some information on the manurial nitrogen values of indoor table chickens fed Label Rouge rations. The dietary specifications of the Label Rouge rations were similar to those of feeds used for organic table chickens, and crude protein, lysine and methionine contents, and the metabolisable energy values (ME) of the Label Rouge rations, are given in Table 13.

Table 13. CRUDE PROTEIN, LYSINE AND METHIONINE CONTENTS (g/kg) AND METABOLISABLE ENERGY VALUES (MJ/kg) OF LABEL ROUGE RATIONS, ON A FRESH BASIS

Ration	Dry matter (g/kg)	Crude protein (g/kg)	Lysine (g/kg)	Methionine (g/kg)	ME value (MJ/kg)
Starter	873.0	182.0	1.03	0.37	12.3
Grower	871.0	161.0	0.86	0.31	12.5
Finisher	870.0	159.0	0.76	0.29	12.5

Source: DEFRA-funded project OF1053

The cereal contents of the Label Rouge starter, grower and finisher rations as a proportion of the ingredients, on a fresh matter basis, were 65.5%, 75.0% and 76.0%. Although synthetic lysine and methionine were used in the Label Rouge rations their inclusions in the diets were at concentrations commensurate with meeting the birds' needs for health and well-being, and not at concentrations aimed at optimising growth. The manurial nitrogen (N) values of indoor table chickens fed either presumed non-limiting rations or Label Rouge rations on an *ad libitum* basis are shown in Tables 14 and 15.

Table 14. TOTAL NITROGEN, AMMONIUM-N AND URIC ACID-N MEASURED IN MANURE SAMPLED FROM INDOOR ROSS 308 BIRDS[1] FED *AD LIBITUM* TO DAY 81 EITHER PRESUMED NON-LIMITING RATIONS (NL) OR LABEL ROUGE RATIONS (LR), COMPARED WITH PUBLISHED VALUES FOR CONVENTIONAL BROILERS (NICHOLSON *et al.*, 1996)

	NL	*LR*	*Nicholson et al., (1996)*
Dry matter (%)	49.7	47.7	64.2
Total N (%dw)	6	5	5
Total N (kg/t fw)	29	23	33
NH_4-N (%dw)	1.8	1.3	1
NH_4-N (kg/t fw)	9	6	6
Uric acid-N (%dw)	0.4	1.6	0.7
Uric acid-N (kg/t fw)	8	2	4

[1]Ross 308 birds are fast growing hybrids widely used in conventional broiler production. Note that to achieve typical market live weights of between 2.0 kg and 2.5 kg at day 81 it would be necessary to restrict daily feed intakes. As Ross 308 birds were fed *ad libitum* the manurial nitrogen values may be higher than in feed restricted birds.

Table 15. TOTAL NITROGEN, AMMONIUM-N AND URIC ACID-N MEASURED IN MANURE SAMPLED FROM INDOOR ISA 657 BIRDS[2] FED *AD LIBITUM* TO DAY 81 EITHER PRESUMED NON-LIMITING RATIONS (NL) OR LABEL ROUGE RATIONS (LR), COMPARED WITH PUBLISHED VALUES FOR CONVENTIONAL BROILERS (NICHOLSON *et al.*, 1996)

	NL	*LR*	*Nicholson et al., (1996)*
Dry matter (%)	65.3	69.9	64.2
Total N (%dw)	5	4	5
Total N (kg/t fw)	33	25	33
NH_4-N (%dw)	0.8	0.5	1
NH_4-N (kg/t fw)	5	3	6
Uric acid-N (%dw)	0.8	0.2	0.7
Uric acid-N (kg/t fw)	5	1	4

[2] ISA 657 birds are widely used in Label Rouge table bird production. They were fed *ad libitum* to day 81 and at this age they had achieved a typical market live weight of about 2.2 kg.

The readily-plant-available nitrogen supply (ammonium-N plus uric acid-N, MAFF, 1994) of manure from Ross 308 birds fed either presumed non-limiting rations or Label Rouge rations was within the range reported by Nicholson *et al.,* (1996) for conventional broilers. For ISA 657 birds, it was similar to the mean value reported by Nicholson *et al.,* (1996) for broilers, when ISA 657 birds were fed presumed non-limiting rations, but it was towards the lower range when fed Label Rouge rations. The dry matter content of litter from ISA 657 birds fed Label Rouge rations was higher than the mean value reported for broiler litters (about 70%, compared with about 60%, respectively), and the total nitrogen content was slightly lower (about 4% on a dry weight basis (dw), compared with about 6% dw). As breed growth profiles and feed intakes will differ it is suggested that in order to optimise the utilisation of manure nutrients, the nutrient content of table bird manures should be checked prior to land application.

MANNER has been used to illustrate the effects of breed on table chicken manure application rates and readily-available-plant-N (Table 16, Chambers *et al.,* 1999). *MANNER* is a DEFRA-funded decision support system that is used to accurately predict the fertiliser nitrogen value of livestock manures on a field specific basis . It allows for leaching and volatilisation losses and takes account of the organic fraction of the manure. Input data to the model are easily obtained: manure analysis (defaults available), rainfall, application date and method, and soil and subsoil texture. Calculations were made using the manure analysis presented in Tables 14 and 15, compared with non-organic broiler litter. Comparisons were made for three scenarios: 1) surface applied on 1 November, left on surface; 2) surface applied on 1 November, ploughed down within three days, and; 3) surface applied on 1 March, ploughed within one day. Table 17 shows the default values set for the *MANNER* model.

Loss pathways and plant available N are driven by the plant available fraction of the manure: the sum of ammonimum and uric acid N. Thus, manures with a low proportion of this compared with the total N will have less short-term availability to the crop and pose less of an envrionmental risk in terms of leaching and volatilisation losses. For this reason, the ISA 657 birds fed a LR ration tended to lose less N.

The example shows that these manures are valuable sources of N for crops: but the phosphorus (P) and potassium (K) should also be taken into account, as many organic farmers may value these nutrients more than the additional nitrogen supplied.

Table 16. ENVIRONMENTAL FATE OF POULTRY LITTER FOR DIFFERENT SCENARIOS ACCORDING TO TIME AND METHOD OF APPLICATION. THESE SCENARIOS ARE: 1) SURFACE APPLIED 1 NOVEMBER AND REMAINING ON SURFACE; 2) SURFACE APPLIED 1 MARCH AND PLOUGHED WITHIN THREE DAYS, AND; 3) SURFACE APPLIED 1 MARCH AND PLOUGHED WITHIN ONE DAY

	Ross 308 birds			*ISA 657 birds*			*'Conventional' litter*		
	(1)	*(2)*	*(3)*	*(1)*	*(2)*	*(3)*	*(1)*	*(2)*	*(3)*
Application rate (t/ha)	7.4	7.4	7.4	6.8	6.8	6.8	5.1	5.1	5.1
N applied (kg/ha)	170	170	170	170	170	170	168	168	168
Potentially available N (kg/ha)	70	70	70	42	42	42	63	63	63
Leached N (kg/ha)	35	50	0	16	23	0	30	43	2
Volatilised N (kg/ha)	24	8	2	11	4	1	21	7	0
Available N (kg/ha)	12	12	68	15	15	40	12	12	61

Table 17. DEFAULT VALUES FOR *MANNER* MODEL USED TO TEST FERTILISER VALUE FOR POULTRY LITTER

Parameter	*Value*
Application rate	To apply 170 kg/ha total N
Top soil (0-30 cm) /Sub soil (30-90 cm)	Loamy sand to depth
Rainfall	Average for Gleadthorpe
Manure analysis	As in Tables 14 and 15, 'standard' value as for non-organic broiler litter

Source: ADAS

10

WHOLE FARM SYSTEMS INCLUDING ORGANIC CHICKENS: NUTRIENT BALANCES AND VIABILITY

Mark Shepherd and Anne Bhogul (ADAS, Gleadthorpe)

A study of the potential place of organic chickens in whole farm systems is not straightforward, but it is essential if progress is to be made in devising such systems. This section of the book considers nutrient cycling between the soil, crops and birds in terms of sustainability and nutrient budgets. Limitations of the nutrient balance approach, and of the viability of the whole farm system in terms of nutrient balance are discussed.

Nutrient cycling and sustainability

Sustainability is an underlying principle of organic farming systems. Nutrients removed in crop and animal products must be replaced. Organic farmers aim to achieve this by the recycling of nutrients on farm and by using natural processes such as nitrogen fixation. They aim to minimise the need for external inputs, and as far as possible to work within a closed system.

To optimise nutrient recycling within a farm, it is important to plan rotations carefully. Organic rotations are characterised by fertility-building (based on legumes) and fertility-depleting phases. This can result in considerable variation in the nutrient status of soils, particularly in relation to N which is more labile, but less so for P and K levels which change more slowly. A well designed rotation will allow for these fluctuations by selecting a sequence of crops with different nutrient requirements to match the changing supplies (Shepherd *et al.*, 2000). N is generally considered the limiting nutrient in many organic rotations. The rotation will typically begin with a fertility building phase which, for stockless systems, will usually comprise of a one or two year ley containing legumes such as red or white clover, vetch or trefoil to fix atmospheric N. Mixed livestock systems use this phase more efficiently by grazing grass/clover leys. The inclusion of organic poultry enterprises within such a system can therefore make good use of the ley and provide additional income. Poultry do not have to be the sole livestock enterprise, as the ley can also be grazed by cattle (Lampkin, 1994). Several years of arable cropping can usually be sustained after this initial fertility building phase, but this will typically include a fertility building crop such as peas or beans midway through the rotation.

NUTRIENT BALANCES

One way of assessing the sustainability of a farming system is to construct a nutrient balance. This is a nutrient accounting process which sums all the inputs and outputs to and from a given, defined system (Watson and Stockdale, 1999; Igras, 2001). An excess of inputs over outputs will result in a surplus, which can lead to an increase in soil fertility, but which may also have environmental consequences such as water or air pollution (e.g. nitrate leaching, and ammonia or nitrous oxide emissions respectively). Conversely, an excess of outputs over inputs will result in a deficit, leading to a progressive decline in fertility which may affect the long-term sustainability of the system. Comparison of nutrient inputs with outputs therefore provides an indication of the likely long term changes in soil nutrient status. Whole-farm nutrient balances (or budgets) are not only a way of assessing the viability of farm rotations, but they can also be used to determine best management practices, to predict the impact of changing practices and to compare systems (Watson and Stockdale, 1999). Nutrient balances form a major part of agricultural policy in the Netherlands (Munters, 1997).

There are a number of approaches to compiling nutrient balances. The simplest identifies the farmgate as the system boundary and treats the farm as a 'black box'. Only nutrients which enter or leave the farm via the 'gate' are considered. Therefore, inputs include purchased fertilisers, manure, feedstuffs, animals and seeds, and outputs include sold plant and animal products (Watson and Stockdale, 1999). Strictly, a 'farmgate' balance will not include inputs in biological nitrogen fixation and atmospheric deposition, although these are sometimes included (Igras, 2001). Loss of nutrients by leaching or gaseous emissions are excluded from this type of balance calculation.

An alternative approach, is the construction of a soil surface balance, which estimates the net loading of nutrients to the soil (Igras, 2001). All nutrients that enter the soil are considered as inputs, including nitrogen fixation, atmospheric deposition, fertilisers, manures and seeds. Besides the output in harvested crops, this method also includes losses as leaching and gaseous emissions from the soil. A soil surface balance therefore excludes the input and output of nutrients in animal feeds and products, as these should be accounted for by animal manures and harvested crops.

CASE STUDIES

As mentioned above, non-organic poultry production has often been separated from the land, and the integration of poultry into organic farming systems presents many technical challenges. Perhaps the most challenging is the balancing of crop and poultry production so that the bulk of the feed can be supplied from home-grown sources, without compromising welfare and environmental objectives.

The integration of chickens into an organic rotation is currently quite rare. Nutrient (N, P, K) balances were therefore constructed in order to assess the viability of integrating poultry production systems within a typical hypothetical organic rotation. This included one or two years of grass/ clover ley as the initial fertility building phase, followed by potatoes and wheat, with a crop of peas to boost soil fertility, and ending with a second wheat crop. Farmgate and soil surface balances were calculated for this rotation, assuming that part of the ley was utilised by either laying hens or table birds. Although the example used here is hypothetical, it is very similar to a commercial rotation identified in DEFRA-funded project OF0178 on improved N management in organic systems, where a seven course rotation is followed, including two years of grass/clover ley (with chickens) followed by one year each of winter wheat or potatoes, oats, winter beans or peas, winter wheat and spring barley (Berry, *personal communication*).

METHODOLOGY

It was assumed that each phase of the hypothetical rotation occupied a 3 ha block. Crop yields were estimated from Lampkin (1990) and Lampkin *et al.,* (1999), with nutrient offtakes calculated from the total dry matter removed and the nutrient content (Lampkin, 1990; Larbier and Leclercq, 1994; Karklins, 2001), (Table 1). This resulted in a total of 84 t potatoes,

27 t wheat and 9 t peas, with 15 t of baled straw (assuming a harvest index of 47%, with 50% of the non-grain biomass removed by baling; Anon, 1997). The nutrient content of harvested products (Table 1) was also used to estimate the nutrient input in seeds and tubers, using typical seed rates: 200 kg wheat seed/ha, 100 kg pea seed/ha, 3 t potato tubers/ha and 28 kg grass/clover seed/ha (Lampkin *et al.*, 1999). Biological nitrogen fixation was estimated to contribute 150 kg/ha N and 105 kg/ha N from the grass/clover ley and peas, respectively (Lampkin, 1990; Wood, 1996, Fisher, 1996), with atmospheric deposition estimated at 20 kg N/ha/yr, 0.22 kg P/ha/yr and 3.32 kg K/ha/yr for lowland England and Wales (Mitchell *et al.*, 1998).

Table 1. DRY MATTER (DM) YIELD AND NUTRIENT OFFTAKE OF CROPS IN THE HYPOTHETICAL ROTATION (3 ha OF EACH CROP)

Rotation position	Crop	Yield (t/ha DM)	Nutrient removal kg/t, DM			kg		
			N	P	K	N	P	K
1	Grass/clover	9[a]	10[a]	1.3[a]	4.2[a]	270	35	112
2	Grass/clover	10[a]	10[a]	1.3[a]	4.2[a]	300	39	124
3	Potatoes	28[b]	2.5[c]	0.4[c]	4.8[c]	210	34	403
4	Wheat	5[b]	18[d]	3.3[d]	4.6[c]	268	49	69
5	Peas	3[b]	33[d]	4.5[d]	11[d]	291	40	99
6	Wheat	4[b]	18[d]	3.3[d]	4.6[c]	215	40	55

Source: [a] Lampkin (1990); [b] Lampkin and Measures (1999); [c] Karklins (2001); [d] Larbier and Leclercq (1994)

Part of the grass/clover ley was assumed to accomodate the chickens. For laying hens UKROFS Standards allow a maximum of 1 000 hens/ha and for table chickens a maximum of 2 500 birds/ha. The hypothetical rotation could therefore accommodate up to 6 000 laying hens or 15 000 table chickens, but the accompanying arable rotation would not support flock sizes of this magnitude (see below). It was calculated that a more realistic flock size would be 1 000 laying hens or 2 500 table chickens occupying just one hectare of the ley, with the remaining ley (5 ha) maintained by cutting and mulching. In practice, mobile housing would probably be used, so that the birds could be moved to different areas of the ley within the flock cycle. The vacant ley could also potentially be used to produce silage or for grazing an additional livestock enterprise.

Laying hens

INPUTS

It is current practice within the UK for non-organic pullets to be transferred from the rearing site to the organic laying unit at 16 weeks of age. The birds are then reared according to organic standards for six weeks, after which they achieve organic status. At this age, 22 weeks, the hens are given access to pasture and eggs laid from then onwards may be sold as organic eggs, as well as meeting the EC marketing criteria for free-range eggs.

Between 16 and 72 weeks of age an organic layer will consume about 50 kg feed. To optimise nutrient recycling within the rotation as much home-grown feedstuffs as possible should be used to feed the hens. The energy component of the ration is not likely to be a problem, except that sufficient quantities of wheat may not be provided by the rotation (see below). The biggest challenge will be in meeting the hens' essential amino acid requirements using home grown protein sources, in particular lysine and methionine. This is because most plant protein sources are deficient in lysine and/or methionine, and in the example rotation only one protein source is produced: peas. A ration based on wheat and peas would not support egg production, and there would be health and welfare implications. Therefore, it was decided to formulate a ration based on current practice within the organic sector, using both soya bean meal and synthetic amino acids (Table 2).

A derogation on synthetic amino acids in poultry feeds on the basis of health and welfare was in place at the time of writing. Furthermore, by using an optimal feed formulation greater confidence is given to the egg mass output data and manure nutrient contents. Inaccuracies in the latter would have affected the accuracy of the nutrient budget calculations. Feeds high in crude protein but low in methionine and lysine would be associated with low egg outputs, according to response to the first limiting amino acid (e.g. McDonald and Morris, 1985).

One flock of 1 000 laying hens could be housed on 1 ha of the grass/clover ley. This flock would require 50 t feed, of which 35 t would be wheat, with the remainder (15 t) bought in from external sources (protein, minerals and vitamins). However, the hypothetical rotation only produces 27 t wheat, so a further 8 t would have to be purchased. The input of N, P and K in purchased feeds (Table 3) was calculated by difference from the dietary N, P and K requirements and known wheat N, P and K contents (Larbier and Leclercq, 1994; Karklins, 2001). In order to supply all the wheat from home-grown sources the rotation could be extended to include a second crop of wheat at the end (as in the example of project OF0178, described earlier) or the field size increased.

Table 2. RATION FORMULATION FOR ORGANIC LAYING HENS

Ingredient	Quantity (kg/t)
Wheat	704.6
Gluten meal	68.4
Full fat soya	98.0
Grassmeal	10.0
Minerals, vitamins and amino acids	119.0

Source: ADAS

Table 3. INPUT OF N, P AND K IN PURCHASED FEEDS (1 000 BIRDS)

Nutrient	Total feed intake (kg)	Wheat content (kg)*	Soya, gluten etc (kg)
N	1316	144 (630)	686
P	267	26 (116)	151
K	254	37 (167)	92

* values in parenthesis are from home-grown sources

The hypothetical rotation would provide ample straw for the flock's bedding requirements, with 1 000 laying hens requiring *circa* 8 m^3 or 4 t straw, assuming 33% of the house is covered (Elson, *personal communication*).

OUTPUTS

Using the ration described in Table 2 the pea crop, grown to boost fertility mid-way through the arable rotation, could be sold (3 t/ha), together with the potatoes (28 t/ha). In addition, a laying hen will typically produce 260 to 310 eggs per hen housed in 52 weeks in lay, and 280 has been assumed for the following calculations, with a crude protein content of 106 g/kg egg or 16.96 g N/kg (Shrimpton, 1987). A whole raw egg (excluding the shell) contains about 200 mg P/100g and 130 mg K/100g (Holland *et al.*, 1991). Assuming a mean egg weight of 59 g , with 9.5% shell containing negligible P and K (Larbier and Leclercq, 1994), a whole raw egg will contain 107 mg P and 69 mg K.

Relatively low egg production and egg weight examples have been chosen for the calculations which follow, for several reasons. Firstly, in some organic enterprises traditional breeds may be used. Secondly, and more generally applicable, technical reasons to expect low performance include feed amino acid balance and constraints on feed ingredients, and lighting regime. Also the calculations below do not allow for mortality, so a low example egg production per surviving bird takes this into account.

The total nutrient output in eggs from a flock of 1000 hens would therefore be 280 kg N, 30 kg P and 19.5 kg K. Laying hens gain very little weight during the flock cycle, so the output of nutrients in the sale of spent hens would tend to balance the input in pullets. The input/output of nutrients in body mass was therefore excluded from the balance calculations. Mortality losses during the flock cycle are difficult to estimate and were also excluded.

A layer will produce about 40 to 45 kg of fresh droppings during the flock cycle. Approximately 60% of this was assumed to be deposited in the housing (about 26 t from 1 000 birds) with the remainder (about 17 t) on the grass/clover ley. Layer manure collected from the housing would probably be applied to the potato crop at a rate of about 8.5 t/ha, and was therefore considered an input to the soil surface balance. The input of nutrients with this application (plus that dropped directly onto the ley) was calculated assuming a manure nutrient content of 16 kg N/t, 5.7 kg P/t and 7.5 kg K/ t (Anon, 2000). These are based on analyses of conventional layer manures including any bedding material, collected either directly from the housing or from the heaps immediately after emptying the houses. As a result, gaseous N losses during the housing period will automatically be included within the balance. Loss of N (by leaching and volatilisation) following manure spreading was estimated using the *MANNER* model (Chambers *et al.*, 1999) assuming that the manure was applied prior to the potato crop in spring, on a light textured soil and incorporated within two days of application. This would result in a loss of about 10 kg N/ha via ammonia volatilisation, but negligible leaching. Denitrification losses were not considered. Volatilisation losses from droppings excreted directly onto the grass/clover ley were estimated assuming 45% of the excreted N was lost (Misselbrook *et al.*, 2000). Loss of nutrients during manure storage was not considered.

Nitrate leaching losses were also considered in the soil surface balance. Results from a three year study conducted by Elm Farm Research Centre suggested about 21 kg N/ha/yr, 75 kg N/ha/yr and 52 kg N/ha/yr would be lost from the ley, first year arable and subsequent arable cropping years respectively (EFRC, 1997). This demonstrates the potential for large losses during the transition from the ley to arable phases of the rotation.

Table chickens

INPUTS

A table bird chick weighing about 38 g contains about 80.8 g N/kg, 10.2 g P/kg and 2.0 g K/kg (DEFRA-funded project SP0119), and between 1 and 81 days of age, it will consume about 6.9 kg feed (DEFRA-funded project OF0153). This was assumed to consist of 0.635 kg/bird starter, 1.63 kg/bird grower and 4.635 kg/bird finisher, using studies at ADAS Gleadthorpe as a guide. As with the laying hens, it was decided to formulate a ration based on current practice within the organic sector, with soya bean meal as the main protein source rather than home grown peas, although these did feature within the ration; Table 4. The nutrient specification of this ration is given in Table 5.

Table 4. RATION FORMULATION USED FOR ORGANIC TABLE CHICKENS

Ingredient	Quantity (kg/t)		
	Starter	*Grower*	*Finisher*
Wheat	550	700	710
Wheatfeed	105	50	50
Full fat soya	260	198	192
Peas	50	20	17
Starter supplement	35	-	-
Grower supplement	-	32	-
Finisher supplement	-	-	30

Source: DEFRA-funded project OF0153

Within a year, four flocks of 2 500 table birds (10 000 birds in total) could be accommodated on 1 ha of the grass/clover ley. They would require 69 t feed (6.3 t starter, 16.3 t grower and 46.4 t finisher), consisting of about 48 t of wheat and 1.5 t peas, with the remainder (19.5 t) bought in from external sources. However, the hypothetical rotation would only produce 27 t wheat, so a further 21 t would have to be purchased, unless the field size was increased, or the rotation extended to accommodate further wheat crops. The input of N, P and K in purchased feeds (Table 6) was calculated by difference from the dietary N, P and K requirements and known wheat and pea N, P and K contents (Larbier and Leclercq, 1994; Karklins, 2001).

Table 5. NUTRIENT SPECIFICATION OF TABLE BIRD RATIONS (AS FED BASIS)

Content (g/kg)	Starter	Grower	Finisher
Dry matter	873	871	870
Crude protein	182	161	159
Nitrogen	29.1	25.8	25.4
Total phosphorus	7.3	6.5	6.5
Potassium	8.0	6.7	6.6

Source: DEFRA-funded project OF1053

Table 6. INPUT OF N, P AND K IN PURCHASED FEEDS (4 FLOCKS OF 2500 BIRDS)

Nutrient	Total feed intake (kg)	Wheat content (kg)*	Pea content (kg)**	Soya, gluten etc (kg)
N	1784	856 (376)	48	880
P	452	156 (68)	8	288
K	464	220 (96)	16	228

* values in parenthesis are from home-grown sources
** supplied from home-grown sources only

It was assumed that the straw requirement would be identical to that suggested for the laying hen case study (i.e. *circa* 8 m^3 or 4 t straw/1000 birds). However, with 2 500 birds/flock, the hypothetical rotation would only supply enough to house 1.5 flocks, with a further 25 t needing to be purchased from external sources.

OUTPUTS

Using the ration described in Table 4, the bulk of the pea crop, grown to boost fertility mid-way through the arable rotation, could be sold (about 7 t), together with the potatoes (84 t). In addition, a maximum of 10 000 birds could be sold each year. Table birds typically contain 195 g/kg crude protein or 31.2 g N/kg (DEFRA-funded project OF0163), 12.1 g P/kg and 2.0 g K/kg (Hurwitz and Plavnik, 1986). At a target weight of 2.2 kg/bird this is equivalent to 68.6 g, 26.6 g and 4.4 g of N, P and K, respectively per bird. As with the laying hens, mortality losses during the flock cycle are difficult to estimate and were assumed to be negligible.

Manure output for table chickens reaching the target weight of 2.2 kg/bird by day 81 was estimated at about 3.4 kg/bird (DEFRA-funded project OF0153). Again, approximately 60% of this was assumed to be deposited in the housing (about 20 t from four flocks of 2 500 birds) with the remainder (about 14 t) on the grass/clover ley. Layer manure collected from the housing would probably be applied to the potato crop at a rate of about 6.5 t/ha, and was therefore considered an input to the soil surface balance. The input of nutrients with this application, plus that dropped directly onto the ley, was calculated assuming a manure nutrient content of 30 kg N/t, 11 kg P/t and 15 kg K/t (Anon, 2000). As with the laying hens, these values are based on analyses of conventional broiler litters, including any bedding material, collected either directly from the housing or from the heaps immediately after emptying the houses. Consequently, gaseous N losses during the housing period were automatically included within the balance. Loss of N, by leaching and volatilisation, following manure spreading was estimated using the *MANNER* model (Chambers *et al.*, 1999), as before assuming the manure was applied prior to the potato crop in spring, on a light textured soil, and incorporated within two days of application. This would result in a loss of about 5 kg/ha N via ammonia volatilisation, but negligible leaching. Volatilisation losses from droppings excreted directly onto the grass/clover ley were again estimated assuming 45% of the excreted N was lost (Misselbrook *et al.*, 2000). Loss of nutrients during manure storage was not considered and nitrate leaching losses were assumed to be identical with those given for the laying hens case study.

Results

Several different scenarios were considered in the construction of farmgate and soil surface nutrient (N, P, K) balances for the rotation described above: a six course rotation (two years grass/clover ley prior to the arable cropping), with laying hens or table birds, compared to a five course rotation (one year grass/clover ley) with laying hens or table birds. The farmgate balances included inputs of nutrients by atmospheric deposition and nitrogen fixation, purchased seeds, animals (table bird chicks only), feeds and bedding materials. These were compared with the output of nutrients in sold plant (potatoes/peas) and animal (eggs/birds) products. The soil surface balances also included nutrient inputs as atmospheric deposition, nitrogen fixation and seeds, but feeds, purchased animals and bedding materials were excluded as these should be accounted for by including animal manures as an input. These inputs were compared with the output of nutrients in all harvested plant products, including those recycled on-farm, nitrate leaching losses and ammonia volatilisation losses following manure spreading.

Table 7. FARMGATE NUTRIENT BALANCE FOR ORGANIC POULTRY MEAT PRODUCTION INTEGRATED WITHIN A SIX COURSE ROTATION (TWO YEARS GRASS/CLOVER, POTATOES, WHEAT, PEAS AND WHEAT; 18 ha)

	Amount of nutrients for the farm		
	N	*P*	*K*
	INPUTS (kg)		
Atmospheric deposition	360	4	60
Biologically fixed nitrogen	1 215	0	0
Purchased seeds and tubers	55	9	53
Purchased rearing animals	31	4	0.8
Purchased animal feeds and bedding	1 381	365	452
TOTAL	*3 032*	*382*	*565.8*
	OUTPUTS (kg)		
Sold plant products	437	65	480
Sold animal products	686	266	44
TOTAL	*1 123*	*331*	*524*
	BALANCE (INPUTS-OUTPUTS)		
Farm balance (kg)	1 919	51	41
Balance (kg/ha)	107	2.8	2.3

Tables 7 and 8 are examples of full farmgate and soil surface nutrient balances for organic table birds integrated within a six course rotation. Results from the other scenarios are given in Table 9. This also includes results from farmgate balances constructed for rotations where all the cereal feed requirements can be supplied by home-grown sources (i.e. by either adding another cereal crop at the end of the rotation or by increasing the field size).

Table 8.   SOIL SURFACE NUTRIENT BALANCE FOR ORGANIC POULTRY MEAT PRODUCTION INTEGRATED WITHIN A SIX COURSE ROTATION (TWO YEARS GRASS/CLOVER, POTATOES, WHEAT, PEAS AND WHEAT; 18 ha).

| | *Amount of nutrients for the farm* | | |
	N	*P*	*K*
	INPUTS (kg)		
Atmospheric deposition	360	4	60
Biologically fixed nitrogen	1215	0	0
Seeds and tubers	55	9	53
Animal manures*	1020	374	510
TOTAL	*2650*	*387*	*623*
	OUTPUTS (kg)		
Harvested crops	984	163	626
Leaching	819	0	0
TOTAL	*1803*	*163*	*626*
	BALANCE (INPUTS-OUTPUTS)		
Soil surface balance (kg)	847	224	-4
Balance (kg/ha)	47	12	-0.2

* Includes correction for loss of N following manure application

The farmgate nutrient budgets were very similar for N and P regardless of the length of rotation or form of chicken production (Table 9). On average, there was a surplus of 85 kg N/ha over all the different scenarios, with the surplus being slightly greater from table chicken production and from the six course rather than five course rotation. This was mainly due to a greater input of N in purchased feeds for the table chickens, as four flocks could be sustained within one year, compared to just one flock of laying hens. Also, two years of grass/clover ley in the six course rotation provided greater potential for N inputs by fixation. The P balance was near break-

Table 9. FARMGATE AND SOIL SURFACE NUTRIENT BALANCES (kg/ha) FOR ORGANIC MEAT AND EGG PRODUCTION INTEGRATED WITHIN TYPICAL FARM ROTATIONS.

	Amount of nutrients for the farm (kg/ha)		
	N	*P*	*K*
	FARMGATE NUTRIENT BALANCE		
Table birds: 6 course rotation (18 ha)	107	2.8	2.3
Table birds: 5 course rotation (15 ha)	94	3.3	2.1
Laying hens: 6 course rotation (18 ha)	93	4.8	-17
Laying hens: 5 course rotation (15 ha)	78	5.7	-19
Alternative scenarios to maximise home-grown cereal feed production:			
Table birds: 7 course rotation (21 ha)	83	0.6	-1.5
Table birds: 6 course rotation (36 ha)	72	-2.3	-14
Laying hens: 7 course rotation (21 ha)	76	3.0	-14
Laying hens: 6 course rotation (24 ha)	80	1.7	-19
	SOIL SURFACE NUTRIENT BALANCE		
Table birds: 6 course rotation (18 ha)	47	12	-0.2
Table birds: 5 course rotation (15 ha)	27	15	-0.9
Laying hens: 6 course rotation (18 ha)	29	5.3	-11
Laying hens: 5 course rotation (15 ha)	5	6.3	-13

even for the farmgate budgets, with a small average surplus of about 2 kg P/ha. This was greater for egg production due to a smaller output of P in sold animal products. There tended to be a deficit of K in all scenarios (average: 10 kg/ha), except for table chicken production where some of the cereals had to be purchased (5 and 6 course rotations). The deficit was largest for eggs, which had a lower input of K in purchased feeds.

There was also a surplus of N in the soil surface balances (Table 8), but it was much lower (average: 27 kg N/ha) than that in the farmgate balances, probably because losses of N via leaching and ammonia volatilisation were included, which reduced the surplus. In contrast, the P surplus was greater in the soil surface balance compared to the farmgate balance because the output of P from harvested crops (considered in the soil surface balance) was much less than that from sold plant and animal products (considered in the farmgate balance). The K deficit was also greater in the soil surface balance compared with the farmgate balance for table chickens, but smaller in the case of laying hens. This was due to greater output of K in harvested crops (soil surface) compared to sold plant and animal products (farmgate) for the table chickens, but a higher input of K in manures (soil surface) compared to purchased feeds (farmgate) for the laying hens.

Limitations to the nutrient balance approach

Nutrient balances provide only a guide to the potential viability of a rotation or farming system, not least because there are errors associated with the assumptions and calculations (crop and manure nutrient contents, for example). The case studies given are subject to a number of errors which could change the overall conclusions, particularly for N which currently suggests the rotations are viable, albeit with a surplus. The main N inputs are from biological N fixation and purchased feeds (or manures in the case of the soil surface balance). N fixation is difficult to measure, particularly on a farm scale, with estimates for grass/clover leys ranging from 100-450 kg/ha (Blake, 1987; Lampkin, 1994; Wood, 1996). The actual amount fixed will depend on a number of factors, including soil type, climate, clover cultivar, proportion of clover in the sward and age of the sward. As fixation forms a major N input into the balance, any variation in the estimates used is likely to have a large impact on the overall balance. Manure nutrient analyses are also notoriously variable (Smith and Chambers, 1995). This, together with the fact that the balances were based on the nutrient content of chicken manures from non-organic farming systems, could also lead to error in the overall soil surface balances. Losses of nutrients during storage of manures were not considered, neither were mortality losses. However, despite these uncertainties, the case studies given provide a best estimate for the likely nutrient budget of chicken production systems integrated within typical organic crop rotations. They therefore provide a useful guide to the likely viability of such systems and highlight potential weaknesses and areas requiring further research.

Whole farm budgets for organic farms, particularly livestock farms, frequently show a N surplus (Shepherd *et al.*, 2000). This is largely due to the contribution of N from fixation, but also purchased feeds. By contrast, P and K budgets for organic farms frequently indicate deficits, as seen in the case studies above. The main P and K sources are from purchased feeds (farmgate) or animal manures (soil surface). It is interesting that the input of P and K in purchased feeds for the table chicken case studies was similar to the input of P and K in manures. This was not the case for N, which was less from the manure input compared to the feed input, probably due to losses during housing and application. For laying hens these relationships were not apparent, due to a lower nutrient content of the manures. A continued surplus of N could lead to an increase in losses, whereas a continued deficit of P and K could lead to a withdrawal of soil reserves. Ensuring an adequate P and K supply is a major challenge to most organic farmers (Stockdale, 1999).

The balances rely heavily on the input of nutrients in purchased feeds. As discussed, the rations used in the case studies were based on current practice within the organic sector, with most of the protein sources purchased from external sources. However, problems with sourcing GM-free, organic soya, together with the possible future requirement of supplying all feedstuffs from home-grown sources, may mean more of the feed will have to be home-grown. This could lead to increased nutrient deficits.

Nutrient balance and the viability of organic poultry systems

The construction of nutrient balances for assessing the viability of organic poultry production systems has demonstrated that a closed system where all the inputs can be supplied on-farm is currently not achievable. Some feeds, particularly protein sources, have to be purchased and these provide a valuable source of nutrients (particularly P and K), which are in deficit on many organic farms. However, although the principle of a closed cycle is fundamental to organic farming, it need not necessarily apply at an individual farm level (Lampkin, 1994). It could be argued that on a regional scale, it does not matter if the feed crops, the poultry and the land upon which the manure is used are in different locations (Charles, 1996). One way forward may be to form co-operative groups, in each of which several organic arable farms supply one organic poultry producer with feed in return for manure.

11

ROTATIONS AND CONSTRAINTS

There is a need to examine how well chickens fit within rotations in terms of pasture availability throughout a typical laying year, or over the minimum growing period. Furthermore, this task must take into account the minimum pasture rest periods as stipulated by UKROFS. This is discussed with reference to the hypothetical rotation used in Chapter 10.

Fitting into the rotation

The hypothetical organic rotation used for calculating nutrient budgets is illustrated in Figure 1. It was a six course rotation consisting of two years of grass/clover ley, followed by one year of potatoes, one year of wheat, one year of peas, and lastly one year of wheat. In this rotation, laying hens or table chickens may graze the grass/clover ley.

Year 1

[1]Grass/ clover	[2]Grass/ Clover
[6]Wheat	[3]Pot's
[5]Peas	[4]Wheat

Year 2

[1]Grass/ clover	[2]Pot's
[6]Grass/ clover	[3]Wheat
[5]Wheat	[4]Peas

Year 3

[1]Pot's	[2]Wheat
[6]Grass/ clover	[3]Peas
[5]Grass/ clover	[4]Wheat

Year 4

[1]Wheat	[2]Peas
[6]Pot's	[3]Wheat
[5]Grass/ clover	[4]Grass/ Clover

Year 5

[1]Peas	[2]Wheat
[6]Wheat	[3]Grass/ clover
[5]Pot's	[4]Grass/

Year 6

[1]Wheat	[2]Grass/ clover
[6]Peas	Grass/ clover
[5]Wheat clover	[4]Pot's

where, pot's = potatoes

Figure 1. Theoretical rotation as used in Chapter 10.

There is a need to examine how well chickens fit within the above rotation in terms of pasture availability throughout a typical laying year, or over the minimum growing period for table chickens. Furthermore, this task must take into account the minimum pasture rest periods stipulated by UKROFS.

The approximate dates for sowing grass/clover swards and the availability of pasture for grazing by poultry on a plot basis are shown in Figure 2. The two year grass/clover ley is available for grazing by chickens for about 21 or 22 months.

Month	Year 1	Year 2	Year 3	Year 4	Year 5	Year 6
Jan	↑	↑	│	│	│	│
	establish		↓			
Feb	grass/		plough			
	clover				sow peas	
Mar	sward		sow			
			potatoes		│	
Apr	↓		│			
May	↑		│			
June	│	│	harvest			
	│	│	if 2nd			
July	grass/	grass/	early	↓	↓	↓
	clover	clover	variety	harvest	harvest	harvest
Aug	sward	sward	↓ or	Wheat	peas	wheat
	- option	- option	harvest			
Sept	graze	graze	if main		sow	sow
	by	by	crop		winter	grass/
Oct	chickens	chickens	variety		wheat	clover
			Sow		│	sward
Nov	│	│	winter		│	│
	│	│	wheat		│	│
Dec	↓	↓	│		↓	↓
			↓			

Figure 2. Theoretical rotation for one block as used in Chapter 10.

Laying hens

It is current practice within the UK for conventional pullets to be transferred from the rearing site to the organic laying unit at 16 weeks of age. The birds are then reared according to organic standards for six weeks, after which time they achieve organic status. At 22 weeks the hens are given access to pasture and eggs laid from then onwards may be sold as organic eggs. They also meet the EC marketing criteria for free-range eggs.

The integration of organic laying hens into the hypothetical rotation used in Chapter 10 and re-iterated earlier in this chapter is illustrated in Figure 3.

Month	Plot 1			Plot 6		
	Year 1	Year 2	Year 3	Year 1	Year 2	Year 3
Jan	↑ establish	grows			↑ establish	
Feb	grass/		plough		grass/	
Mar	clover sward		sow potatoes		clover sward	
Apr						
May	↓ ↑	↓			↓	
June	laying hens		harvest if 2nd			
July	given		early variety	↓	↑ house	
Aug	access to pasture		↓ or harvest	harvest wheat	pullets at 16 weeks	↓
Sept	at 22 weeks		if main crop	sow	of age access to	
Oct	of age (housed		variety Sow	grass/ clover	pasture given	
Nov	at 16 weeks		winter wheat	sward	at 22 weeks	
Dec	of age whilst grass		↓	↓	of age	

Figure 3. Theoretical grazing of a grass/clover ley by laying hens housed at 16 weeks of age, given access to pasture at 22 weeks of age and depopulated at 72 weeks of age - as used in Chapter 10.

In the above example, slightly more than one year of a grass/clover ley is occupied by laying hens. If, as in Chapter 10, a plot is 3 ha and the flock size is 1 000 laying hens, only 1 ha or less (depending on sector body standards) of grass/clover ley would be needed per flock. This means that the flock may be rotated around the 3 ha of pasture so as to spread the manure and parasites.

At the end of the flock the hens would be depopulated and the houses cleaned. The houses would then be moved to the next plot occupied by year 1 of the grass/clover ley. Note that the production period used causes slippage, and so as the rotation progresses laying hens may increasingly occupy year 2 of the grass/clover ley.

If producers decide to restock the same pasture then a two month rest period would be required. In practice this would mean that a two week cleanout period is optimal in terms of "fitting the rotation", followed by restocking with 16 week-old non-organic pullets. The pullets would then undergo a six week conversion period to organic status; at the end of the conversion period the hens would be 22 weeks of age. Outdoor access would be allowed at 22 weeks of age, as the pasture would have been rested for two months and the eggs may be sold as organic. However, it would not be possible to complete the second flock of laying hens on the same pasture as the grass/clover ley must be ploughed-in so as allow potatoes to be sown before the hens reach 72 weeks of age. The practicalities of moving houses from one plot to another when fully stocked, and the likely deleterious effects on bird well-being, performance and egg quality must be considered.

There is a derogation in EC 1804/1999 on the rearing of organic pullets and this expires after the 31 December 2003. If the derogation is not extended then this means that there will be a need to rear pullets in an organic system, and possibly with access to pasture. The latter would give rise to many different scenarios, but it is likely pullet rearing would be done in different facilities and on different land from that used by laying hens. One reason for this is that nest boxes would not be needed in the rearing houses provided that the pullets are moved to the laying site prior to laying their first egg. Nest boxes in the rearing house would reduce the available stocking area, and therefore income per unit floor area. However, a more important reason for segregating the rearing and laying phase is disease control, including parasites.

Table chickens

The minimum slaughter age for organic table chickens is 81 days of age unless a recognised slow growing breed is used. In the latter case, non-organic chicks may be bought in to an organic system before three days of age and then reared according to organic standards for a period of 70 days before gaining organic status. In the examples given below an 81-day growing period has been used. This is because it illustrates the minimum number of flocks of table chickens possible within the hypothetical rotation defined above.

In the example given below, year 1 only of the grass/clover ley is used for grazing by table chickens. If as in Chapter 10 a plot is 3 ha and the flock size is 2 500 table chickens, only 1 ha of grass/clover would be needed per flock. This means that each flock could be rotated within the 3 ha so as to spread the manure and parasites. Thus, the same pasture would be used once only except for 1 ha, and this would be used twice, but with a rest period of about six months. This is three times the requirement for pasture rest (UKROFS, 2001).

An alternative approach would be to have seven flocks of 2500 birds graze each two year grass/clover ley. This would maximise the number of flocks produced for the hypothetical rotation given in Chapter 10, and it is illustrated in Figure 5. The birds would be rotated within the 3 ha of pasture and this would give a pasture rest period of about 6 months for each 1 ha of land. However, this approach would incur extra capital costs as additional housing would be needed for concurrent flocks and there would be extra labour. It would also mean that when using the hypothetical rotation given in Chapter 10 there would be a larger deficit of on-farm cereals, straw and protein for use in the table chicken enterprise. Pasture would be more heavily utilised and this may impact on the manure and parasite build-up on land.

A further option would be to brood chicks centrally until about 28 days of age, followed by transfer to free-range facilities for the remainder of the growing period. If birds are to have access to range soon after housing (e.g. at 29 days of age) and a cleanout period of 14 days is used, then each 1 ha of pasture would have a rest period of about 4.5 months before being restocked. If, however, birds are housed at 28 days of age, but not given access to pasture until 36 days of age, then the cleanout period may be reduced to seven days.

Month	Plot 1			Plot 6		
	Year 1	Year 2	Year 3	Year 1	Year 2	Year 3
Jan	↑ establish grass/ clover sward	↓			↑ Establish grass/ Clover Sward	↓
Feb		↑	plough			↑
Mar		Flock 4	sow			Flock 8
			potatoes			
Apr	|	↓	|	|	|	|
May	↓↑	↓			↓↑	↓
June	Flock 1		↓ harvest		Flock 5	
			if 2nd			
July			early	↓		
Aug	↓↑		variety ↓ or harvest	harvest wheat	↓↑	
Sept	Flock 2		if main	sow	Flock 6	
			crop	grass/		
Oct	↓		variety Sow	clover	↓	
Nov	↑		winter wheat	sward	↑	
Dec	Flock 3		| ↓	↓	Flock 7	

Figure 4. Hypothetical grazing of a grass/clover ley by table chickens grown to 81 days of age and including a cleanout period of about 13 days - as used in Chapter 10.

Month	Plot 1			Plot 6		
	Year 1	Year 2	Year 3	Year 1	Year 2	Year 3
Jan	↑	∣	∣	∣	↑	∣
Feb	establish	↓	plough	∣	establish	↓
	grass/	↑		∣	grass/	↑
Mar	clover sward	∣ Flock 4	sow	∣	clover sward	∣ Flock 11
			potatoes	∣		
Apr	∣	∣	∣	∣	∣	∣
May	↓ ↑	↓ ↑	∣	∣	↓ ↑	↓ ↑
June	∣ Flock 1	Flock 5	↓ harvest	∣	∣ Flock 6	∣ Flock 12
			if 2nd	∣		
July	∣	∣	early	↓	↓	∣
Aug	↓ ↑	↓ ↑	variety ↓ or	harvest wheat	↑	↓ ↑
	∣	∣	harvest		∣	∣
Sept	Flock 2	Flock 7	if main	sow	Flock 8	Flock 14
			crop	grass/		
Oct	∣	∣	variety	clover	∣	∣
	↓	↓	Sow	sward	↓	↓
Nov	↑	↑	winter	∣	↑	↑
	∣	∣	wheat	∣	∣	∣
Dec	Flock 3	Flock 9	∣	∣	Flock 10	Flock 16
			↓	↓		

Figure 5. Hypothetical grazing of a grass/clover ley by table chickens grown to 81 days of age and including a cleanout period of about 13 days

Month	Plot 1			Plot 6		
	Year 1	Year 2	Year 3	Year 1	Year 2	Year 3
Jan	↑ Establish	Flock 4			↑ establish	Flock 9
Feb	Grass/ Clover	↓	plough		grass/ clover	↓
Mar	Sward	↑ Flock 5	sow potatoes		sward	↑ Flock 10
Apr		↓				↓
May	↑ Flock 1				↓ ↑ Flock 6	
June			harvest			
July	↓ ↑		if 2nd early variety	↓	↓ ↑	
Aug	Flock 2		↓ or	harvest wheat	Flock 7	
Sept	↓		harvest if main	sow	↓	
Oct	↑ Flock 3		crop variety Sow	grass/ clover sward	↑ Flock 8	
Nov			winter			
Dec	↓ ↑		wheat ↓	↓	↓ ↑	

Figure 6. Hypothetical grazing of a grass/clover ley by table chickens placed in free-range facilities at about 28 days of age after centralised brooding, grown to 81 days of age and including a cleanout period of about 14 days

The availability of capital and labour

The integration of organic chickens into whole farm rotations will depend upon: farm activities throughout the year, the availability of labour and stockman skills, capital expenditure and investment in facilities, and disease status. For example, it may be not be desirable to have two flocks of table chickens running concurrently during the summer months (Figure 5) during harvesting, because at that time labour may be a limiting factor.

Centralised brooding may be a sensible option for both the poultry industry and for producers wanting to diversify by adding a second or third enterprise to an organic farm. It would allow specialist husbandry skills to be applied during a crucial period of chick development. In addition, investment in brooding facilities, including environment control, failsafes and alarms would be more feasible, compared with providing similar equipment in several small houses having no fixed location on the farm. For example, alarms for high or low temperatures in the brooding houses would be of little use if they are not audible for most of the working day and during the night.

However, a balance between stocking age, access to range and the cleanout period is needed. An approach based on maximising the number of flocks *per annum* may not be acceptable to the organic consumer if table chickens are given access to range for a relatively short period of their life. A frequent stocking schedule and a short cleanout period is likely to increase the risk of disease carry over, and this must be discussed with the site veterinarian. The minimum cleanout period may need to be extended if there has been a disease outbreak.

Consideration must be given to the methods used for catching, containing and transporting birds when transferring from brooding to growing facilities. This event is stressful to chickens and there is the potential for injury. Trauma from unobserved injuries will reduce bird welfare and they may be apparent only at processing. It is important to take into account the impact of season on welfare during catching and transport. There may be heat stress when catching and moving birds in hot weather and the job may be best done during the cooler part of the day but before darkness. Conversely, during winter birds may become cold stressed, particularly in cold, wet and windy weather.

Discussion

The key ethos of organic farming is the concept of the soil as a living system, and that there is an essential link between soil, plants, animals and man. Nutrients should not be taken from the soil in the form of feed ingredients for livestock if nutrients are not going to be added back to the soil through manure.

However, it would not be fair to assume that there has been no consideration of nutrient recycling in non-organic systems of poultry production. In the conventional period, drying has produced high nutrient value manures and these have been used in arable rotations. In addition, the responses to protein and essential amino acid intake have been widely studied and the findings have helped to reduce the overfeeding of these nutrients. The chicken excretes excess nitrogen and the use of high N manure on land may lead to pollution. Recycling of nutrients sometimes occurred through the processing and feeding of waste products such as feathers, but this is no longer permitted in the EU.

An aim of this discussion is to consider how well current systems of organic chicken production meet the ethos of organic farming, and how this may change in the future. The discussion is based around the principles and practices of organic farming systems according to the standards of the International Federation of Organic Agricultural Movements (IFOAM) given below:

- to produce food of high nutritional quality in sufficient quantity
- to work with natural systems rather than seeking to dominate them
- to encourage and enhance biological cycles within the farming system, involving micro-organisms, soil flora and fauna, plants and animals
- to maintain and increase the long-term fertility of soils
- to use as far as possible renewable sources in locally organised agricultural systems
- to work as much as possible within a closed system with regard to organic matter and nutrient elements
- to give all livestock conditions of life that allow them to perform all aspects of innate behaviour
- to avoid all forms of pollution that may result from agricultural techniques

- to maintain the genetic diversity of the agricultural system and its surroundings, including the protection of plant and wildlife habitats
- to allow agricultural producers an adequate return and satisfaction from their work including a safe working environment, and
- to consider the wider social and ecological impact of the farming system.

Consumer demand for organic poultry products, and in particular for eggs, has been high in the UK. At present, the supply of UK organic eggs is thought to meet demand, but whether or not more consumers would choose to buy organic eggs if the price were lower is not known. If consumer demand for organic eggs were to become entirely price led then there is a risk of failure to meet IFOAM standards. When derogations on stocking density and pullet rearing end the cost of organic egg production will increase. This may price organic eggs out of the market for some UK consumers.

There is currently a deficit between consumer demand for organic chicken meat and UK production. This means that organic chicken meat is being imported into the UK. The social and environmental impact of doing this needs to be studied.

One reason cited for the deficit between demand for organic chicken meat and production is the need for 'slow growing' breeds. The growing period in organic production is almost twice as long as that for broilers, but the required market live weight is similar between systems. This means that broiler hybrids grown in an organic system are too heavy at the end of the growing period unless they have been feed restricted. Feed restriction is not thought to be a suitable technique for use in organic systems when the basis is cost. Broiler hybrid chicks are widely available in the UK; whereas 'slow growing' hybrids were, until recently, only available by imports from France, either as hatching eggs or chicks. Thus, the unit cost of 'slow growing' chicks is greater than that of broiler hybrid chicks.

A lack of information on breed performance and meat yields may have delayed expansion in the organic chicken meat sector, but this has been addressed by DEFRA-funded research (Project OF0153) and by industry in-house research.

If technical difficulties at some times of the year (e.g. pasture management, or the application of light programmes) lead to seasonal production in the UK, it is likely that consumer demand will be met through imports. Methods of storing eggs and chicken meat over prolonged periods and without detriment to the product may be important, but demand is most likely to be for fresh produce. Some parts of Southern Europe may be better able to avoid seasonal production and these countries would be the likely source of 'out-of-season' organic chicken products. However, the social and environmental impact of importing organic eggs and chicken meat at specific times of the year would need to be studied. In terms of cost to the environment it is likely to be preferable to have long-term storage of home-produced products than to transport products over long distances.

Generally, the nutrient content of organic eggs or poultry meat is not expected to differ from that of non-organic eggs or poultry meat. The protein content of eggs or meat is thought to be fairly consistent. This is because proteins have a biological role within the egg or bird. There will be more body fat in organic chickens than in broilers because fat accumulates with age. In addition, in an outdoor system the environmental temperature will fluctuate widely and this will make it harder to balance the supply of feed energy and protein. If birds over-consume feed energy in order to meet their protein requirements this will lead to fat deposition.

Organic rations are likely to comprise a wide ingredient base, and if there is a reliance on oilseed by-products then the fatty acid composition of both the feed and product may differ to that of non-organic. This is because the fatty acid composition of eggs and chicken meat is readily altered through dietary means. This was the subject of a review by Gordon (2001).

The fatty acid composition of a lipid is important: it affects the quality of the food during storage. Quality is poor when food is rancid and when this occurs it is rejected because of poor taste. Poor taste is due to the presence of ketones and aldehydes; these being the end products of fatty acid oxidation.

If rations are rich in unsaturated fatty acids there may be a need to increase the concentration of antioxidants in the feed, but antioxidant choice is important as it needs to be effective in the bird (e.g. α-tocopherol acetate). There is information available suggesting that some herbs may have an antioxidant role in chicken meat, (reviewed by Gordon 2001), and provided that processing methods are consistent with the methods of organic production they may be useful ingredients.

'Functional' foods are foods thought to provide some health benefits when eaten in sufficient quantities. There is an increasing market for 'functional' foods and the egg industry diversified their product range so as to enter this market. The product is an omega-3 enriched egg. To produce this, hens are fed rations rich in omega-3 fatty acids and this may be achieved through the feeding of fish oil, fish meal or marine microalgae. Although omega-3 rich eggs are niche products at present, they may one day be the norm. If organic egg producers are to be able to diversify their product range so as to include omega-3 rich eggs, then fish by-products will need to be permitted feed ingredients.

Aside from nutrient value, the safety of food products in terms of bacterial load needs to be considered. There have been substantial efforts in the poultry industry to reduce the load of food poisoning bacteria in the products. Some of the control strategies employed cannot be applied in organic production systems. For example, the poultry building forms a physical barrier to vermin and wild birds in indoor systems and this prevents the feed from being contaminated. In outdoor systems, the house is open to vermin and wild birds and their droppings may easily contaminate the pasture. Feeding outdoors may encourage chickens to range, although the distance ranged is often limited to where feed provision ends, and the feed is likely to attract wild birds. Methods of encouraging ranging other than feeding outdoors will be important in organic poultry production, and in particular if high nutrient value swards are identified and used.

Natural methods of increasing the birds' resistance to food poisoning bacteria are needed. In view of this, feed ingredients having antibiotic properties, such as herbs, are likely to be important. The efficacy of herbs as antibiotics in feeds for poultry should be determined.

Methods for reducing the risk of crops being contaminated by food poisoning bacteria present in non-organic poultry manure have been identified (FSA project BOS003,). Batch storage of manure and allowing time between application of manure and harvest were the important factors. However, there is a need to examine whether or not there are differences in the bacterial population and bacterial loads of organic poultry manures, compared with non-organic manures. This may be more important for poultry grazing land previously occupied by other livestock. Cattle excrete *E.coli* 0157 in their manure, whereas to date poultry have not been identified as a source of this very serious food poisoning bacterium.

Several of the IFOAM standards given above are inter-related, and so in the section below they are discussed together.

Some organic producers have integrated their poultry enterprise into their cropping system (FRFR) and to do this they use mobile houses. The houses are moved onto new pasture as the rotation progresses. This method of integration is consistent with the ethos of organic production. Manuring of the pasture occurs during ranging and parasite control may be better achieved by relatively infrequent occupation of the pasture compared with fixed housing systems. However, there has been little consideration of the role of pasture as a food source for the birds. Grass/clover leys are usually used: they are hard wearing and the clover is important because of its role in fertility building. In organic systems the pasture should contribute to the birds' nutrient intake, otherwise nutrients in the sward are wasted and there is an over dependence on bought-in feed.

For several reasons, chickens in the conventional period were separated from the land. This means that there is little information available about grass intake in chickens and its nutrient value to the bird. There is a need to develop swards suited for use by poultry, and this may mean a move away from grass-based swards to alternative forages rich in protein and some essential amino acids. This approach may reduce the requirement for bought-in protein.

The effect of cropping system and sward composition on soil invertebrate populations should be studied: insects living in the soil surface may be a high protein food source for poultry, but they may also control plant pests and diseases. A drop in some invertebrate populations, because chickens are eating them, may make natural means of plant disease control less effective. Thus, it will be necessary to study the effects of incorporating poultry into cropping systems on invertebrate diversity and population.

Tree canopies provide shaded areas outdoors and this is well used by chickens. The birds tend to sit in the shaded areas, whereas in other parts of the pasture standing is predominant. In FRFR systems it may be more difficult to provide birds with shade from tree canopies apart from on the perimeter of the field. This is because trees dotted throughout the field would impede sowing, harvesting and the growth of crops. As chickens have a preference for shade and shelter outdoors it would seem sensible to combine organic poultry production with fruit production or agroforestry.

The effects of incorporating organic poultry into cropping systems on nutrient budgets, ecosystems, and the biological control of disease in plants and chickens need to be studied. If grass/clover leys are not effectively utilised by poultry then other livestock may graze them and this may be beneficial in the cropping system in terms of nutrient recycling. The grass/clover ley is likely to remain an important fertility building phase in the rotation, and in future cropping systems poultry may occur at another phase in the rotation.

For the hypothetical rotation used in the calculation of nutrient budgets (two years of grass/clover ley, followed by one year each of potatoes, wheat, peas, and lastly wheat) at best only the cereal component of the feed was home-produced. For this rotation, a nitrogen surplus was found, but only because protein was bought-in. Phosphorus balance was near breakeven, and there was a deficit of potassium.

It is likely that the rotation will need to be expanded somewhat in order to integrate chickens better. This is because of increasing pressure to produce more feed on-farm and to move towards a closed system. In the hypothetical rotation, the only protein source produced for the chickens was peas. It is unlikely that any one home-produced protein source will meet the birds' requirements for essential amino acids. All of the vegetable-based protein sources reviewed by Gordon (1999), and in Chapter 3, were deficient in one or more essential amino acid and they contain antinutritive factors that limited their inclusion rate. Thus, a wider range of proteinaceous crops will be need to be included in the rotation. Even then, the birds' essential amino acid requirements are not likely to be met solely through the use of home-produced ingredients. This means that there may be a reliance on bought-in proteinaceous ingredients rich in essential amino acids, including fish meal.

There is a need to formulate rations using as much as possible of the home-produced ingredients. The rations should have sufficient nutrients for health and welfare, and they should support an acceptable level of performance. The rations must not lead to pollution because of the necessity of feeding an excess of some nutrients. It is likely that it will be most difficult to avoid feeding an excess of nitrogen as feed crude protein may need to be increased in order to meet the birds' essential amino acid requirements.

Once rations have been formulated, the quantities of home-produced ingredients needed per unit of feed will be known. This information may then be used to devise hypothetical rotations capable of producing most of the cereal and protein components of the feed. The viability of the rotations will then need to be tested by calculating nutrient budgets. For the most promising hypothetical rotations, field-based studies will need to be done and, in addition to monitoring nutrient flow through the system, the impact on ecosystems should be studied.

The amount of feed eaten by outdoor poultry will vary with the time of year and outdoor temperature. This will affect the quantities of feed ingredients that need to be produced.

Feeds are used less efficiently by chickens when housed in extensive systems because outdoor temperatures in the UK cause higher levels of body heat loss at most times of the year than those occurring in indoor systems. Feather cover is important in reducing heat loss and there may be some breed differences in the rate of feathering and the quality of feathering. Some traditional breeds such as Orpingtons have feathering extending low down on their legs and this may reduce heat loss when standing or walking. Whether or not leg feathers become dirty in organic production systems would need to be assessed: wet and dirty feathers will be ineffective in reducing heat loss.

If cold temperatures are predicted then the ratio of feed energy value to protein of the ration should be widened so as to avoid high feed intakes and an unnecessarily high protein intake. However, it may be more difficult to increase the feed energy value of organic rations because the high cereal content reduces space within the ration for other ingredients. Oils have a high energy value and they readily increase the energy value of the feed without occupying a great deal of space, but it is not easy to source organic oils for use in livestock feed and price is an issue. If oils cannot be used to increase the energy content of the ration then full-fat oilseeds may be need to be used.

Insulating the house so as to reduce house heat loss at night when the popholes are closed may improve feed conversion. A sustainable way of doing this may be to stack straw bales on the outside of the building, but not so as to block air inlets and outlets. The microbiological quality of the straw would need to be good so as to avoid disease problems. Furthermore, the rotation would need to provide sufficient straw and it would be needed for bedding as well as insulation.

Vermin control would need to be good so as to avoid infestation in the house outer straw layer. This is easier to achieve in fixed paddocks as the perimeter fence may extend below the soil surface so as to prevent vermin burrowing beneath the fence. It would be labour intensive to do the same for FRFR systems.

Shelter outdoors provides beneficial microclimates and these may be inhabited by the chickens. If they do use the microclimate then heat loss will be lower, and in particular if the birds are sitting on dry pasture. Straw bale tunnels underneath tree canopies may be attractive to the birds and effective in reducing heat loss. Conifer wig-wams, as used in project OF0153, may also provide a microclimate for the birds, but this would depend on leaf cover.

There are nutrients lost from the system when poultry products are taken off the farm and sold. Using human excreta in organic cropping systems would be one means of recycling some of the nutrients from organic foods, but this is not acceptable because of disease risks. Nutrients are also lost from the system when birds die. Composting of dead birds would be one means of recycling the nutrients, but this is not allowed. If the organic chicken meat market expands greatly then there may be increasing pressure to find acceptable means of recycling or utilising some of the waste products of processing. For example, whether or not chicken feathers may be processed and used, eg. perhaps as an insulation material in livestock buildings, is one such possibility. If the latter were possible then this would increase environmental friendliness.

One of the standards of organic production given by IFOAM is that livestock should be allowed to perform all aspects of innate behaviour. Extensive systems of poultry production may provide the best opportunity for this, but none of the current systems fully meet this standard. That is because reproduction remains as a separate aspect of production. Table chickens are slaughtered just prior to sexual development, and so breeding should not occur. In organic egg production, laying hens are not allowed to breed: the eggs are for human consumption. Furthermore, in modern layer hybrids the tendency towards broodiness has been much reduced. Breeding is done by having dedicated breeder flocks and the eggs are incubated in modern hi-tech hatcheries and not by the hen. This approach is not likely to change because there would be implications for meat quality if birds were slaughtered after breeding, and eggs would be larger but fewer from older hens. There would be substantial economic costs associated with a less controlled breeding and production system.

An issue that will be important when the derogation on organic pullet rearing expires at the end of December 2003, is the utilisation of the males of egg lines. In non-organic production the males of egg lines are killed at hatching as they are not useful in meat production, because their growth profile and conformation are poor compared with modern broiler hybrids. This practice will also be applied in organic breeder and pullet production, as male chicks of egg lines will not perform well enough in organic table chicken production; meat yields will not be acceptable. This raises an ethical issue: the killing of organic male chicks of egg lines because they do not have a useful purpose in production.

The above situation suggests that dual-purpose breeds may be of use in future organic systems. The male chicks may be used for meat production and the female chicks may be reared for egg production. This would be a solution, but performance is expected to be poorer than for breeds developed for a specific purpose. Thus, the costs of production would be greater and this would need to be met by the consumer.

Some traditional breeds are dual-purpose and their use in organic production is encouraged by the standards. However, the cost of production is critical because if costs are too high the retail price will inhibit sales and expansion of the organic poultry sector.

An exercise by Turner (2001) illustrated that when the derogation on stocking density for organic laying hens ends, gross margins will be negative unless egg prices increase significantly. Turner (2001) considered mobile housing in his calculations and the egg enterprise was fully integrated into the cropping system. Using less efficient breeds for egg production would make the situation worse.

The organic egg sector may face difficulties in coming years as there is an established market for organic eggs in which egg prices are relatively stable. Existing consumers may not be able or willing to tolerate a significant increase in egg price and so consumer demand may fall. If this is the case, it was perhaps unfortunate to allow less stringent standards to be used than those coming into force within the near future. Consumers may feel cheated by price increases and they may not understand why they are necessary. Furthermore, they may not have been aware of the current practice, that is to allow pullets up to 18 weeks of age to be bought-in from non-organic sources and converted to organic within only six weeks.

In summary, future systems of organic poultry production are likely to be much more complex. The poultry enterprise is likely to be fully integrated into the cropping system and the rotations used will be wider than those typical today. Poultry may not be the only livestock on the farm. They poorly utilise grass/clover leys, compared with ruminants, and the grass/clover ley is likely to remain an important fertility building phase. This means that there will be an increased emphasis on identifying sward compositions having a high nutrient value when grazed by poultry.

On-farm feed production will need to increase and this will improve sustainability at a local level. The contribution of home-grown protein sources to the birds' protein requirements will have to increase. Complementary to this would be the use of high protein forages for grazing by poultry.

Poultry manure is a valuable plant nutrient source, but the amount of protein fed will affect the manure nitrogen content. Thus, the fertiliser nitrogen value of the manure should be determined so as to avoid pollution.

The long-term viability of cropping systems including poultry will depend upon the balance of nutrients within the system and the availability of organic matter. Nutrient budgets will need to be calculated for hypothetical rotations that include poultry and most of the feed.

The effects of integrating poultry into cropping systems on ecosystems need to be studied and the implications for plant disease control need to be considered.

Conclusions

Currently, organic table chickens and eggs are produced in fixed or mobile houses, and in the latter case grazing may move around the whole farm (FRFR). There are advantages and disadvantages for each of the above systems and they are detailed below.

ADVANTAGES AND DISADVANTAGES OF FIXED AND MOBILE HOUSING SYSTEMS

Fixed houses usually have good access roads for the delivery of stock and feed, and for the collection of eggs and birds. In FRFR systems access is likely to be limited and the transport of stock, feed and eggs around the site may be dependent on using a tractor and trailer. This will be more labour intensive, and an effective bird transport system will need to be developed so as to avoid poor welfare due to re-handling, climatic stress (wind, rain, heat stress and cold stress) and transport over uneven terrain. An effective system for transporting eggs to a cold store will also be needed or eggs will be cracked during the journey. Eggs should be protected from exposure to sunlight and rain for reasons of quality control. In fixed housing systems the egg store is conveniently located and the movement of eggs is not usually an issue.

For fixed houses, permanent electricity and water supplies may be used, whereas this is not likely to be an option for FRFR systems. The advantage of a mains electricity supply is that it allows a wide range of equipment to be used. For example, automation of air inlets and outlets, feeders and nest boxes, and lighting linked to time clocks and alarms may be used. A permanent water supply is useful, as the water does not freeze. In mobile housing systems water is often stored in a tank and, despite lagging, because of the low volume it may freeze in cold weather.

Fixed houses are often located near to the stockman's house. This means that alarms for high or low temperatures may be used, and in particular during brooding when ambient temperatures must not fluctuate.

The location of fixed houses may be optimised in terms of access and land-type. For example, on a sloping site the house may be located at the highest point and drainage will then be away from the house. In FRFR systems the houses will be rotated around the site and this may mean that the house is sometimes moved to less favourable areas. In fixed housing systems, drainage may be used on heavier land so as to prevent areas of the paddock from becoming waterlogged. Puddles should be avoided as chickens will drink from them and they may be a source of disease-causing organisms.

Fencing may be extended below ground in fixed housing systems and this prevents vermin from burrowing and accessing the house where they may contaminate the feed. This is likely to be too labour intensive in FRFR systems.

In fixed paddocks, it may be easier to incorporate trees whereas in FRFR systems trees may impede sowing, harvesting and crop growth. The shade from tree canopies is useful in encouraging chickens to range.

The control of parasitic disease is expected to be most effective in FRFR systems. This is because FRFR systems are expected to have more land available for grazing by chickens and the period of rest will be longer than in fixed grazing systems. However, there should be some consideration of the use of poultry manures on land to be grazed by chickens. The application of fresh manure on land soon to be grazed by chickens may be a source of contamination.

As the demand for home-produced feed increases rotations will need to be widened. The rotation will need to provide wheat and a variety of proteinaceous ingredients for use in the feed. No single protein source is thought to be suitable for meeting all of the birds' protein and essential amino acid requirements for health and performance. This means that either: a) organic farms will be larger than those typical today so as to accommodate a wider rotation; b) field size may need to be variable; smaller for some crops and larger for other crops so as to balance yields with needs, or; c) co-operatives may need to be formed for trading nutrients at a local level. The third option will be most important for organic poultry producers on small farms.

It is likely that extra capital expenditure will be needed if the number of crops grown is increased. If the new crops are used only for poultry then capital expenditure should be attributed to the poultry enterprise. This is likely to increase the price of organic eggs or organic chicken meat to the consumer.

A narrow hypothetical rotation including chickens (two years of grass/ clover ley followed by one year each of potatoes, wheat, peas, and wheat) that relied heavily on bought-in proteinaceous ingredients produced a farmgate nitrogen surplus, whereas phosphorus was near break-even and potassium was in deficit. A soil surface balance for the same rotation produced a greater surplus of nitrogen, a surplus of phosphorus and a greater deficit of potassium than in the farmgate balance. The long-term viability of this rotation in terms of potassium was not good and the nitrogen surplus may cause pollution.

Poultry are thought to poorly utilise ryegrass-based swards in terms of deriving nutrients from them. In future systems there will be an increased need for the pasture to provide some of the bird's nutrients and this may mean a move away from ryegrass-based swards. Some high protein forages may be of multiple use as a sward, as a proteinaceous feed and as aerial cover for the birds.

If the grass/clover ley is not well utilised by poultry then ruminants may most effectively graze it. The grass/clover ley is important in fertility building and so it is likely to be part of wide rotations.

Nutrient budgets for rotations including poultry, with or without other livestock, will need to be calculated, as this will provide an indication of long-term viability. The calculations will be complex, as it will need to take into account several scenarios, including the level at which it is possible for systems to be closed.

Poultry manure will often contain food poisoning bacteria and so it should be used fresh on land only with great caution. Heaping in batches reduces bacterial loads in poultry manures and there are recommendations on how to do this. There are also recommendations concerning which crops poultry manures may more appropriately be used. Typically these are low risk crops such as cereals.

Heat generated during heaping may kill some poultry worm eggs present in the manure. If so, this technique may be of use in reducing the contamination of land. In FRFR systems poultry graze land that has previously received poultry manure and this may increase worm burdens unless worm eggs are reduced in the manure.

In future systems, there may be a need to use dual-purpose traditional breeds. This is because the current practice of killing male chicks from egg-lines is unlikely to be acceptable in an organic system. If dual-purpose breeds were to be used in organic systems, the males may be used in meat production and the females may be used for egg production. The performance of dual-purpose breeds is likely to be poorer than that of modern egg or meat producing hybrids, and so the cost of production would be greater.

In future systems, where poultry are better integrated into cropping programmes, there may be a risk of eggs being unavailable at some times of the year. Pasture may not be available for grazing and this 'down-time' may be used for cleaning out. The chances of this happening are greater if few cropping systems include laying hens, and if in these systems there is a common time-of-year for establishing the pasture. Agroforestry systems including poultry may not be as susceptible to a seasonal down-time as arable cropping systems. Thus, a wide range of cropping systems including organic poultry should be encouraged so as to avoid seasonal peaks and troughs in egg supply.

In future systems, there will be a greater need for sustainability. This means that the reliance on buying-in poultry houses, and bedding should be reduced. In agroforestry systems the wood could conceivably be used for building poultry houses. In arable cropping systems, straw bales may be used to build houses and if appropriately rendered these are resistant to weathering and well insulated. The weight of straw bales may rule out mobile housing, but this would be worth examining. Rendered houses built of straw are likely to be much cheaper to produce than buying-in modular houses.

Chickens are foragers and they eat insects living in the soil surface. If they eat enough predatory insects so as to affect the ecosystem then natural means of plant disease control may be reduced.

There is a need to identify feeds rich in protein and essential amino acids that may be used in organic poultry production. Novel feeds such, as insect-based feeds should not be ignored as they may have something to offer. Methods of cultivation and feeding would need to be developed, and these would have to be acceptable to consumers.

Integrating organic poultry into cropping systems will provide an extra income, but as a minimum, fixed and variable costs associated with the poultry enterprise need to be met. Recent EU legislation (EC 1804/1999) may significantly impact on egg income when derogations end. Most of the cost of organic egg production is due to capital expenditure and depreciation on housing and feed costs. If these costs can be reduced then gross margins will be better, and any necessary increases in organic egg prices will be smaller.

There may be a need for additional labour on farms and for new skills. Employment opportunities in rural areas may be limited and organic farming may potentially increase the scope for employment in local economies.

Organic eggs account for a large part of the UK organic food sector in terms of product value. There are socio-economic considerations beyond the egg producer as many people are employed throughout the UK organic food supply-chain and in UKROFS approved accreditation bodies.

References – Chapters 9, 10 and 11

Ackert, J.E. (1931). The morpholgy and life history of the fowl nematode *Ascardia lineata* (Schneider). *Parasitology* **23:**360-379

Ambrosen, T. (1996). The Danish Poultry Council. *Personal comunication* with Permin *et al .,* (1999)

Anon (1997). *The Wheat Growth Guide.* Home Grown Cereals Authority, London

Anon (2000). *Fertiliser Recommendations for Agricultural and Horticultural Crops.* RB209. Ministry of Agriculture, Fisheries and Food. London.

Blake, F. (1987). *The Handbook of Organic Husbandry.* Crowood Press, UK

Bray, T.S. and Lancaster, M.B. (1992). The parasitic status of land used by free-range hens. *British Poultry Science* **33:**1119-1120

Carlile, F.S. (1984). Ammonia in poultry houses: a literature review. *World's Poultry Science Journal* **40:**99-113

Chambers, B.J., Lord, E.I., Nicholson, F.A. and Smith, K.A. (1999). Predicting nitrogen availability and losses following application of organic manures to arable land: *MANNER. Soil use and management* **15:**137-143

Charles, D.R. (1984). A model of egg production. *British Poultry Science* **25:** 309-321

Charles, D.R. (1994). Comparative climatic requirements. In: *Livestock housing*. Eds. Wathes, C.M. and Charles, D.R., CAB International, Wallingford, 3-24

Charles, D.R. (1996). *The poultry industry, the environment and the rural economy*. 5th Temperton Fellowship Report. Harper Adams Agricultural College

Charles, D.R., Elson, H.A. and Haywood, M.P.S. (1994). Poultry housing. In: *Livestock housing*. Eds. Wathes, C.M. and Charles, D.R., CAB International, Wallingford, 249-272

Charles, D.R., Groom, C.M. and Bray, T.S. (1981). The effects of temperature on broilers: interactions between temperature and feeding regime. *British Poultry Science* **22:** 475-482

Charles, D.R. and Walker, A.W. (2002) *Poultry environment problems: a guide to their solutions*. Nottingham University Press.

Clark, J.A. and McArthur, A.J. (1994). Thermal exchanges. In: *Livestock housing*. Eds. Wathes, C.M. and Charles, D.R., CAB International, Wallingford, pp97-122

Council Regulation 1907/90 (1990) (amended) on certain Marketing Standards for eggs

Council Directive 1999/74/EC (1999). Laying down minimum standards for the protection of laying hens. This is to be implemented here by the Welfare of Farmed Animals (England) Amendment Regulations 2002 (currently in draft, expected to go before parliament in June 2002)

DEFRA-funded project LS3501. Desk study – Optimising poultry meat quality including eating quality in extensive production systems

DEFRA-funded project OF0153. Effect of breed suitability, system design and management on the welfare and performance of traditional meat birds

DEFRA-funded project OF0163. Optimising the synergism between organic poultry production and whole farm rotations, including home grown protein sources

EC 1804/1999 (1999). Council Regulation supplementing Regulation (EEC) No 2092/91 on organic production of agricultural products and indications referring thereto on agricultural products and foodstuffs to include livestock production

Edge, S.J.A. (2000). Improving litter quality and environment in free-range laying units. MAFF funded project AW0222

EFRC (1997). *Nitrate leaching from organic farming systems.* Report for MAFF Contract NTO 305 (CSA 2461) Elm Farm Research Centre.

Egg (Marketing Standards) Regulation (1995) which implements EU regulations on Marketing Standards

Elson, H.A. (1995). Environmental factors and reproduction. In: *Poultry production. World Animal Science Series.* Ed. Hunton, P., Elsevier, pp389-409

Emmans, G.C. and Charles, D.R. (1977). Climatic environment and poultry feeding in practice. In: *Nutrition and the climatic environment.* Eds. Haresign, W., Swan, H. and Lewis, D., Butterworths, London, pp31-49

Emmans and Fisher, C. (1986). Problems in nutritional theory. In: *Nutrient requirements of poultry and nutritional research.* Eds. Fisher, C. and Boorman, K.N., Butterworths, pp9-39

Fisher, N.M. (1996). The potential of grain and forage legumes in mixed farming systems. In. *Legumes in Sustainable Farming Systems.* Occasional Symposium No. 30 British Grassland Society Ed. Younie, D. pp290-299

Freeman, B.M., Manning, A.C.C. and Flack, I.H. (1980). Short-term stressor effects of food withdrawal on the immature fowl. *Comparative Biochemistry and Physiology* **67A:**569-571

Freeman, B.M., Manning, A.C.C. and Flack, I.H. (1981). The effects of restricted feeding on adrenal corticol activity in the immature fowl. *British Poultry Science* **22:**295-303

FSA-funded project BOS003. Pathogen survival during manure storage and following land application

Gordon, R.F. and Jordan, F.T.W. (1982). *Poultry diseases.* Bailliere and Tindall, London

Gordon, S.H. (1999). The use of home grown protein sources in organic poultry rations. DEFRA-funded project OF0163

Gordon, S.H. (2000). Effect of growing period, breed, management practices and feeding strategies on the quality of extensively produced poultry meat, including organoleptic properties and nutrient content. DEFRA-funded project LS3501

Gordon, S.H. (2001). Effects of modifying the polyunsaturated fatty acid and antioxidant contents of poultry meat on flavour, tenderness and shelf life and its application to extensive systems. DEFRA-funded project LS3501

Holland, B., Welch, A.A., Unwin, I.D., Buss, D.H., Paul, A.A. & Southgate, D.A.T. (1991). McCance and Widdowson's *The composition of foods*, Royal Society of Chemistry and MAFF

Homidan, A. al, Robertson, J.F. and Petchey, A.M. (1996). Some factors affecting dust and ammonia production in broiler houses. In: *World's Poultry Science Association (UK Branch) Proceedings of Spring Meeting*, pp118

Hurwitz, S. and Plavnik, I. (1986). Carcass minerals in chickens (Gallus domesticus) during growth. *Comparative Biochemical Physiology* **83A:**225-227

IFOAM (2000). Basic Standards for Organic Production and Processing

Igbasan, F.A. and Guenter, W. (1996a). The enhancement of the nutritive value of peas for broiler chickens: an evaluation of micronisation and dehulling processes. *Poultry Science* **75:**1243-1252

Igras, J. (2001). A model for nutrient balance calculation. In: *Fertilizers and Fertilisation: Sustainable Nutrient Management, MAINTAIN Report 2*. Eds. M. Fotyma & M. Shepherd. Polish Fertilizer Society - CIEC, 2001, pp75-80

Jordan, F.T.W. and Pattison, M. (1996). *Poultry diseases*. Saunders, London

Karklins, A. (2001). Model for the calculation of nutrient offtake by crop 'OFFTAKE'. In: *Fertilizers and Fertilisation: Sustainable Nutrient Management, MAINTAIN Report 2*. Eds. Fotyma, M. and Shepherd, M. Polish Fertilizer Society - CIEC, 2001 pp63-74

Karunajeewa, H. (1987). A review of current poultry feeding systems and their potential acceptability to animal welfarists. *World's Poultry Science Journal* **43:**20-32

Knott, C.I.F. (2000). Contribution to MAFF project OF0192

Lampkin, N. (1990). *Organic Farming*. Farming Press, Ipswich

Lampkin, N. (1994). *Organic Farming*. Farming Press, Ipswich

Lampkin, N. (1997). *Organic poultry production*. Report to MAFF, CSA 3699

Lampkin, N., Measures, M. and Powell, D. (1999). *1999 Organic farm management handbook*. Welsh Institute of Rural Studies and Elm Farm Research Centre

Larbier, M. and Leclercq, B. (1994). Translated by Wiseman, J. *Nutrition and feeding of poultry*. Nottingham University Press and INRA

Leeson, S., Diaz, G. and Summers, J.D. (1995). *Poultry metabolic disorders and mycotoxins*. University Books, Guelph

Lehninger, A.L., Nelson, D.L. and Cox, M.M. (1993). *Principles of biochemistry*. Worth Publishers, New York, pp511

Lewis, P.D., Perry, G.C., Farmer, L.J. and Patterson, R.L.S. (1997). Responses of two genotypes of chicken to the diets and stocking densities typical of UK and 'Label Rouge' Production systems: I Performance, behaviour and carcass composition. *Meat Science* **45:**501-516

Lund, E.E., Wehr, E.E. and Ellis, D.J. (1966). Earthworm transmission of Heterakis and Histomonas to turkeys and chickens. *Journal of Parasitology* **52:**899-902

MAFF-funded project OF0192. Workshop and desk study to appraise technical difficulties associated with organic pullet rearing

Madsen, H. (1952). A study of nematodes of Danish gallinaecous game-birds. *Danish Review of Game Biology* **1:**126

McDonald, P., Edwards, R.A., Greenhalgh, J.F.D. and Morgan, C.A. (1995). *Animal nutrition.* Longman Scientific and Technical, Harlow

McDonald, P. and Morris, T.R. (1985). Quantitative review of optimum amino acid intakes for young laying pullets. *British Poultry Science* **26:** 253-264

Mead, G.C. (1998). *The safety of poultry products: present trends and future developments.* 7th Temperton Fellowship Report. Harper Adams Agricultural College

Ministry of Agriculture, Fisheries and Food (1983). *Table egg storage on the farm.* Leaflet 563

Ministry of Agriculture, Fisheries and Food (1994). *Fertiliser recommendations.* 6th edition. Booklet RB209, The Stationery Office, London

Ministry of Agriculture, Fisheries and Food (1998). *Code of good agricultural practice.* The air code. MAFF publications, London

Ministry of Agriculture, Fisheries and Food (1998). *Code of good agricultural practice.* The soil code. MAFF publications, London

Ministry of Agriculture, Fisheries and Food (1998). *Code of good agricultural practice.* The water code. MAFF publications, London

Misselbrook, T.H., van der Weerden, T.J., Pain, B.F., Jarvis, S.C., Chambers, B.J., Smith, K.A., Phillips, V.R. and Demmer, T.C.M. (2000). An inventory of Ammonia Emissions from UK Agriculture. *Atmospheric Environment,* **34**, 871-880

Mitchell, R., Chambers, B.J., Webb, J. & Garwood, T. (1998). *Nutrient balances for managed grasslands in England and Wales (1969-93).* Report prepared for MAFF contract CSA2845

Munters, P.J.A.L. (1997). *The Dutch Manure Policy: MINAS (Nutrient Accounting System)*. Report for the Dutch Ministry of Agriculture, Nature Management and Fisheries, pp3

National Research Council (NRC, 1994). *Nutrient requirements for poultry*. Ninth revised edition. National Academy Press, Washington

Nicholson, F.A., Chambers, B.J. and Smith, K.A. (1996). Nutrient composition of poultry manures in England and Wales. *Bioresource Technology* **58**:279-284

Nordenfors, H. and Hoglund, J. (2000). Long term dynamics of *Dermanyssus gallinae* in relation to mite control measure in aviary systems for layers. *British Poultry Science*, **41**:533-540

Owen, J.E. (1994). Structures and materials. In: *Livestock housing*. Eds. Wathes, C.M. and Charles, D.R., CAB International, Wallingford, pp183-246

Payne, C.G. (1966). Environmental temperature and egg production. In: *The physiology of the domestic fowl*. Eds. Horton-Smith, C. and Amoroso, E.C., Oliver and Boyd, Edinburgh, pp235-241

Permin, A., Bisgaard, M., Frandsen, F., Pearman, M., Kold, J. and Nansen, P. (1999). Prevalence of gastrointestinal helminths in different poultry production systems. *British Poultry Science* **40**:439-443

Permin, A. and Nansen, P. (1996). Sygdomsmaesige problemer I den okologiske fjerkrae-produktion med saerlig henblik pa parasitter. In: *okologisk egproduktion*. Bereetning nr. 729 fra Statens Husdyrsbrugsforsog, pp91-97

Permin, A., Nansen, P., Bisgaard, M. and Frandsen, F. (1998). Ascaridia galli infections in free-range layers fed on diets with different protein contents. *British Poultry Science* **39**:441-445

Robinson, L. (1948). *Modern poultry husbandry*. Crosby Lockwood, London

Romijn, C. and Lokhorst, W. (1966). Heat regulation and energy metabolism in the domestic fowl. In: *Physiology of the domestic fowl*. Eds. Horton-Smith, C. and Amoroso, E.C., Oliver and Boyd, Edinburgh, pp211-227

Savory, J. (1976). Broiler growth and feeding behaviour in three different lighting regimes. *British Poultry Science* **17**:557-560

Shepherd, M.A., Harrison, R., Cuttle, S., Johnson, B., Shannon, D. Gosling, P. & Rayns, F. (2000). *Understanding Soil Fertility in Organically Farmed Soils*. A review produced for the Ministry of Agriculture, Fisheries and Food, as part of the contract OF0164. Available from www.adas.co.uk/soilfert

Shrimpton, D.H. (1987). The nutritive value of eggs and their dietary significance. In: *Egg quality - current problems and recent advances.* Eds. Wells, R.G. and Belyavin, C.G., Buterworths, London, pp11-26

Siegel, P.B. and Guhl, A.M. (1956). The measurement of some diurnal rhythms in the activity of White Leghorn Cockerels. *Poultry Science* **35**:1340-1345

Smith, L.P. (1984). *The agricultural climate of England and Wales.* MAFF Reference Book 435

Soulsby, E.J.L. (1982). *Helminths, Arthropods and Protozoa of Domesticated Animals,* Bailliere Tindall, London, pp809

Squance, E. (1966). Protein digestion and metabolism in the colostomised laying hen. In: *Physiology of the domestic fowl.* Eds. Horton-Smith, C. and Amoroso, E.C., Oliver and Boyd, Edinburgh, pp146-154

Stockdale, E.A. (1999). Phosphorus and Potassium Cycling. In: *Organic Farming: Implications for the Environment.* Society of Chemical Industry, 23rd March 1999, London

Stopes, C., Duxbury, R. and Graham, R. (2000). UK organic poultry - what do UK consumers expect? In: *Proceedings 13th International IFOAM Scientific Conference,* Eds. Alfoldi, T., Lockeretz, W.and Niggli, U., pp357

Sturkie, P.D. (1976). *Avian physiology.* Springer-Verlag, New York, pp281

Takeuchi, M. (2000). The mortality of organic chicken. In: *Proceedings 13th International IFOAM Scientific Conference*, Eds. Alfoldi, T., Lockeretz, W. and Niggli, U., pp372

Tucker, S.A. and Walker, A.W. (1992). Hock burn in broilers. In: *Recent advances in animal nutrition.* Eds. Garnsworthy, P.C., Haresign, W. and Cole, D.J.A., Butterworth Heinemann, Oxford, pp33-50

Turner, A.W.B. (2001). An analysis of the effect of introducing poultry into an organic rotation on profitability. DEFRA-funded project OF0163

UKROFS Standards (2001). Standards applying to livestock and livestock products from the following species: bovine (including *Bubalus* and *Bison* species), porcine, ovine, caprine, equidae, poultry. Chapter 1B

Valentine, H. (1964). A study of the effect of different ventilation rates on the ammonia concentrations in the atmosphere of broiler houses. *British Poultry Science* **5**:149-160

Wakelin, D. (1965). Experimental studies on the biology of *Capillaria obsignata*, Madsen 1945, a nematode parasite of the domestic fowl. *Journal of Helminthology* **39:**399-412

Wathes, C.M. and Clark, J.A. (1981a). Sensible heat transfer from the fowl. I. Boundary layer resistance of model fowl. *British Poultry Science* **22:**161-173

Wathes, C.M. and Clark, J.A. (1981b). Sensible heat transfer from the fowl. II. Thermal resistance of the pelt. *British Poultry Science* **22:**175-183

Welfare of Animals (Transport) Order (1997). MAFF publication PB2531 Summary of Law Relating to Farm Animal Welfare

Wheeler and Mayes (1997). *Regional climates of the British Isles*

Wood, M. (1996). Nitrogen Fixation: How much and at what cost? In: *Legumes in Sustainable Farming Systems.* Occasional Symposium No. 30 British Grassland Society Ed. Younie, D., pp26-35

Wray, C. and Wray, A. (2000). *Salmonella in domestic animals.* CABI Publishing, Wallingford

Yahav, S., Straschnov, A., Plavnik, I. and Hurwitz, S. (1996). Effects of diurnally cycling *versus* constant temperatures on chicken growth and food intake. *British Poultry Science* **37:**43-54

INDEX

exercise effects, 100
photoperiodic effects, 82-83, 100
sex effects, 84
temperature effects, 79-82, 98
Growth profiles, 56, 63, 83, 152, 153
Gut flora, 201

HACCP, viii (see in full below)
Harper Adams Agricultural College, 10
Harrowing, 176, 259
Hazard Analysis Critical Control Point
    (HACCP), viii, 209, 250
Helminths, 245
Herbage, nutrient value of, 179
Herbal supplements, 242-243
Heterakis, 244, 246
Historical periods of the poultry
    industry, v
    conventional, v
    historic, vi
    traditional, v
Holdings, numbers of, 4
Household budget, poultry products
    in, 13
Housing, 203-205, 240-241
    mobile, 179, 204, 224
    static, 205, 224
Hubbard ISA, 117, 236
Humidity, 240
Hygiene
    management practice, 208
    pasture management effects, 207
    principles of, 207
    rotation, 247
Hypothesis, x

I457, 117
I657, 75, 117, 135, 231, 253
Insulation, house, 298
International Federation of Organic
    Agricultural Movements (IFOAM),
    292
    standards, IFOAM, 292, 296, 298, 299
Invertebrate populations, 178, 296

Ixworth, 75, 77, 237
Ketones, 140, 294

Labelling of products, 147
Label Rouge, 64, 65, 75, 112, 135, 147,
    233, 262
Labour, availability of, 291
Lactobacillus, 201, 202
Lactones, 140
Layer, laying hen, definitions of, vii
Light intensity, 206
Lighting, role in egg supply
    continuity, 10
Light Sussex, 75, 237
Linolenic acid, 38, 47, 127, 128
Linseed, 37-39
    fatty acids in, 38, 128
    PUFAs in, 39
Litter moisture, 240
Lupins, 30-34
    antinutritional factors in, 31
    inclusion rates of, 32
Lycopene, 41

Maillard reaction, 140
MANNER, 223, 225, 264, 265, 274
Manure, poultry, 224 (see also Poultry
    manure, below)
    output of, 238-239
Market background, 13-18
Market sensitivity, 17
Master Gris, 75, 117, 236
Maturity, degree of, vi
Meadow grass, smooth, 174
Meal eating, 258
Meat, chewiness of, 134
Meat, cohesiveness of, 134
Meat, colour of, 120-123, 154
    age effects, 122
    breed effects, 121-122
    myoglobin, 121-122
    pH, 121
    shelf life, 121, 123